The Ecological Plot

Under the Sign of Nature: Explorations in Environmental Humanities
Sarah Dimick, Alison Glassie, and Jesse Oak Taylor, Editors

The Ecological Plot

HOW STORIES GAVE RISE TO A SCIENCE

John MacNeill Miller

UNIVERSITY OF VIRGINIA PRESS
CHARLOTTESVILLE AND LONDON

The University of Virginia Press is situated on the traditional lands of the Monacan Nation, and the Commonwealth of Virginia was and is home to many other Indigenous people. We pay our respect to all of them, past and present. We also honor the enslaved African and African American people who built the University of Virginia, and we recognize their descendants. We commit to fostering voices from these communities through our publications and to deepening our collective understanding of their histories and contributions.

University of Virginia Press
© 2024 by the Rector and Visitors of the University of Virginia
All rights reserved
Printed in the United States of America on acid-free paper

First published 2024

9 8 7 6 5 4 3 2 1

Library of Congress Cataloging-in-Publication Data

Names: Miller, John MacNeill, author.
Title: The ecological plot : how stories gave rise to a science / John MacNeill Miller.
Description: Charlottesville : University of Virginia Press, 2024. | Series: Under the sign of nature: explorations in environmental humanities | Includes bibliographical references and index.
Identifiers: LCCN 2024000307 (print) | LCCN 2024000308 (ebook) | ISBN 9780813951775 (hardcover) | ISBN 9780813951782 (paperback) | ISBN 9780813951799 (ebook)
Subjects: LCSH: Nature in literature. | Ecology in literature. | Ecocriticism. | English literature—History and criticism.
Classification: LCC PR143 .M55 2024 (print) | LCC PR143 (ebook) | DDC 820.936—dc23/eng/20240228
LC record available at https://lccn.loc.gov/2024000307
LC ebook record available at https://lccn.loc.gov/2024000308

Cover art: From *Mira calligraphiae monumenta*, Georg Bocskay, 1561–62, illuminated by Joris Hoefnagel, 1591–96. (Digital image courtesy of Getty's Open Content Program)
Cover design: Cecilia Sorochin

For my parents,
who never said "no" to a book

Contents

Acknowledgments ix

Introduction: Telling Connections 1
1. Approaching Ecology 17
2. The Price of Interconnection 42
3. The Nature of Fiction 78
4. The Story of Ecology 113

Conclusion: Distant Relations 141

Notes 163
Bibliography 185
Index 201

Acknowledgments

This is a book about how difficult it is to recognize the sprawling, interdependent communities that support us in our everyday lives on this planet. In one way or another, the thinkers discussed here all realize that being alive means racking up a series of debts that extend to so many other people and things that any attempt to account for them all inspires a kind of awe that borders on horror. I find myself facing just that sort of awe and horror now. For well over a decade, some version of this book has been part of my life and thoughts—and that means that a staggering number of people and institutions have been crucial to supporting it. I hope I can do justice to at least a few of them here.

The research and writing of this book were supported by an ACLS Fellowship from the American Council of Learned Societies, which allowed me to step away from teaching at a crucial moment in order to bring the manuscript to fruition.

Far earlier in its life cycle, the work that became this book was made possible by a Mellon Dissertation Fellowship from the School of Arts and Sciences at Rutgers funded through the generosity of the Andrew W. Mellon Foundation. When I was first developing this project, a Graduate Fellowship at the Center for Cultural Analysis at Rutgers and a Qualls Dissertation Fellowship also enabled me to think widely and deeply about the subjects covered here.

Versions of two of the chapters in this book first appeared as articles in academic journals. Portions of chapter 1 were published in *Victorian Literature and Culture* in 2020 as "The Ecological Plot: A Brief History of Multispecies Storytelling, from Malthus to *Middlemarch*." I am grateful to Cambridge University Press for allowing me to reprint a revised version of that essay here. Portions of chapter 4 appeared in *Texas Studies in Literature and Language* in 2020 as "Mischaracterizing the Environment: Hardy,

Darwin, and the Art of Ecological Storytelling." I give thanks to the University of Texas Press for the permission to reprint an altered version of that article in this book.

The dedicated editors and staff members at the University of Virginia Press have done much to help shepherd this manuscript through to publication. I am especially grateful to Angie Hogan for believing in this project from our first correspondence, as well as to the two anonymous readers whose thoughtful, generous feedback dramatically improved the manuscript. Laura Reed-Morrisson caught many mistakes I missed and helped polish this work into its final form.

Jonah Siegel at Rutgers taught me many things—most importantly, how to take scholarship seriously without taking myself too seriously as a scholar. I am so grateful for that lesson. John Kucich and David Kurnick have proven invaluable sources of feedback and support time and again; Amanda Claybaugh, too, offered me crucial direction. Marianne DeKoven's enthusiasm for the nonhuman angle of the project was an inspiration from the start. A number of faculty at Rutgers, especially Lynn Festa, William Galperin, and Colin Jager, taught me how to revise scholarship effectively. Others taught me how to teach, modeling what it means to communicate ideas in a clear and engaging way. Chief among them all was Barry Qualls, an indispensable mentor, champion, and friend. I also owe thanks to Greg Jackson, who showed me it was fun to fight for divisive opinions in front of a crowd. Even before my time at Rutgers, I benefited from a slate of wonderful teachers that included Dan Breen, David Brock, Daniel Christian, Joe Duncan, Thomas Ferraro, Esther Gabara, Cathy Miles, Vin Nardizzi, Jerry Thornbery, and Marianna Torgovnik. Thank you all for your generous teachings.

The collegial environment of the English department at Allegheny College has done much to shape this book. I owe a profound debt to the students, to the English department chairs, and to my fellow teacher-scholars at Allegheny, especially Christopher Bakken, Matthew Ferrence, D. Alexis Hart, Ben Haywood, Jennifer Hellwarth, Doug Jurs, and Jennie Votava. The staff at Pelletier Library—Bill Burlingame, Cynthia Burton, and Linda Ernst in particular—have been models of patience and efficiency in their delivery of materials that made my research and teaching go as smoothly as possible. I was also fortunate to have the research assistance of two students, Madeline Hernstrom-Hill and Anastasia Ripkin, whose work was supported

by a grant from the Andrew W. Mellon Foundation for Collaborative Undergraduate Research in the Humanities at Allegheny.

Outside of Allegheny, I have been incredibly lucky to fall in with the inspiring scholars of literature and ecology affiliated with the Vcologies collective. It would be impossible to thank every Vcologist adequately here, but I'm exceptionally grateful for the support and mentorship of Deanna Kreisel, Allen MacDuffie, Elizabeth Carolyn Miller, Jesse Oak Taylor, and Lynn Voskuil. There are also a number of colleagues across the academy—some Vcologists, some not—whose friendship and camaraderie at conferences over the years have sharpened my wits and made academia a welcoming place to be. Among the many who deserve mention are Mark DiGiacomo, Devin (Roff) Garofalo, Devin (Riff) Griffiths, Priyanka Jacob, Maha Jafri, Jacob Jewusiak, Amanda Kotch, Naomi Levine, Caolan Madden, Ruth McAdams, Kyle McAuley, Alexandra Milsom, Brian Pietras, and Tobias Wilson-Bates.

Of course, I could never have pursued this line of thinking in the first place without the deeper, steadier love and support of those much closer to home. From my early years, my parents and my brother have encouraged whatever weird flights of fancy I had, including my unexpected decision to switch from pursuing biology to studying English literature. I owe so much to the decades of love, curiosity, and conversation they continue to provide. More recently my oldest friend, Ajay Kurian, stepped in with timely encouragement to finish this project when I was thinking of abandoning it. I thank him for that.

Finally, this book and this life would be unthinkable without my wife, Ashley Conlan Miller, and our children, Hugh and Felix. The love, joy, and sustenance they provide is so fundamental and so necessary that it is hard to find the words to thank them. Rather than trying to express it in my own flawed way, then, I will fall back on the wisdom of George Eliot: thank you for "the new life into which I have entered in loving you. . . . [In] knowing you, in loving you, I have had, and still have, what reconciles me to life."

The Ecological Plot

INTRODUCTION

Telling Connections

This book offers a new literary history of ecology.

That description may sound odd, or even like a contradiction in terms. After all, literary history explores how stories, poems, plays, and other creative writings have changed over time. Ecology isn't one of these subjective, expressive kinds of writing. It's a science. Any science worthy of the name should be objective and data-driven. How could a science like ecology have a literary history?

This urge to distinguish between art and science feels so natural that it has almost become instinctive. In reality, though, the separation of the arts and sciences is a recent development. In the early 1800s, when this history begins, "scientist" was not even a word in English, much less a coherent professional identity. People who studied the nonhuman world were known as natural philosophers, natural historians, or simply naturalists. These investigators of nature often dabbled in fiction, nonfiction, and poetry, using a wide variety of methods to explore their ideas and communicate them to the general public. Boundaries between art and science were flexible—if they were acknowledged at all. Distinctions between the natural sciences and the social sciences were virtually unheard-of.[1]

The lack of uniform standards for expertise in these fields had its downsides, to be sure. But it also created a hothouse environment for knowledge exchange in the nineteenth century, opening up possibilities for cross-disciplinary thought that seem lavishly eccentric today. This was an age when the son of a carpenter could be educated in mathematics, accept a faculty position in mineralogy, then transition a few years later to a Chair of Moral Philosophy at Cambridge University. It was an age when a literary critic could independently study current models of the nervous system and use them to write the first book on physiology in English—one of the books that would inspire future Nobel laureate Ivan Pavlov to pursue science. And it was an age when bestselling fiction was widely admired for its ability to reveal overlooked truths about the workings of nature and society alike.[2]

Historically, then, it makes sense to study the origins of a modern science through a literary lens. But using a literary lens to study science makes philosophical sense, too. After all, thought cannot enter the world until it is represented somehow, and scientific thought is no exception. We need words, numbers, graphs, metaphors, and other ways of materializing information before we can transform our first mental inklings into full-fledged ideas. These representations shape our internal experiences into concepts that are stable, coherent, and capable of being shared. Finding the right language or equation or model to embody an idea actually realizes and clarifies that idea. Using less-than-perfect words or equations or models, on the other hand, obscures an idea, leading us to overlook its consequences or ignore the ways it fails to explain reality.

Psychologists and designers have a name for this relationship between ideas and their manifestations in the world. When an object tends to encourage certain ideas or behaviors—the way the gentle dimple atop a button invites pressing, for example—researchers call that inanimate encouragement an "affordance." The button in question *affords* pressing: it makes pressing the button seem like an easy, natural thing to do. But affordances are not just the product of three-dimensional objects like buttons, handles, and chairs. Two-dimensional words and pictures afford thoughts and behaviors, too. They promote certain kinds of thinking and feeling, and they discourage other kinds. The right word or image or sentence can help us come to a conclusion we might otherwise have left unthought. The opposite is also true: when we lack the visual or verbal language to identify something, we struggle to even conceive of its existence. Thus psychologists and anthropologists have found that cultures with more specific words for different shades of blue can immediately isolate and identify those shades in a way that other cultures with a less nuanced vocabulary cannot. When we confront aspects of reality we do not have words for, we experience what psychologists call *hypocognition:* a mental inability to name and process information that lies right in front of us.[3]

What all this means in practice is that ideas are never truly independent of the ways in which we represent them. A representation gives an idea its material form. That form is integral to the idea's realization, development, and movement through the world, just as the bodily form of a person is integral to their constitution, development, and movement. It just so happens that many ideas achieve their clearest, most compelling forms when they are realized through language—including the stories, symbols, and other verbal structures that make up the focus of literary study.

This literary study focuses on one particular historical relationship between an idea and its realized form. The idea is the idea of ecology itself. In the broadest sense, ecology's basic insight—what the scientist and environmentalist Barry Commoner called "The First Law of Ecology"—is the recognition that "Everything Is Connected to Everything Else."[4] More specifically, ecology works to map out the material relationships that connect living things to one another and to their nonliving surroundings. These relationships are based on physical interactions that result in exchanges of matter and energy. Aldo Leopold—the forester, game manager, and conservationist often credited with founding modern environmental ethics—concisely captured the ecological outlook on the world in his influential book *A Sand County Almanac* (1949). There, he explains how "[p]lants absorb energy from the sun. This energy flows through a circuit called the biota." The living system that results entails "lines of dependency for food and other services" that knit lands, waters, plants, animals, and other organisms together into "a sustained circuit, like a slowly augmented revolving fund of life."[5] This ecological understanding of the world is considered scientific. Nevertheless, it came to the attention of modern science only after a mode of storytelling emerged that made it conceivable, comprehensible, and transmissible—a narrative structure I call *the ecological plot*.

That, at least, is the contention of this book. Like other forms of plotting, the ecological plot works by laying out cause-and-effect relationships as they unfold over time.[6] The ecological plot is unique, however, in its attention to the cause-and-effect relationships that Leopold describes in his account of ecology. It dwells on transfers of matter and energy that take place when humans, animals, and plants interact with each other and with their surroundings, the exchanges that connect them all together into a single interdependent network. The staggering complexity of such networks has meant that scientists themselves were relatively slow to arrive at the study of ecology. As Commoner observes, "We have been trained by modern science to think about events that are vastly more simple—how one particle bounces off another, or how molecule A reacts with molecule B."[7] Ecological plots refuse this controlled, isolated simplicity. They imaginatively trace the far-flung consequences of seemingly isolated, local interactions as their effects ripple through much more extended networks. An ecological plot might demonstrate how one decision to withhold money, or one individual's brush with illness, or one pair's erotic encounter, or any other exchange of matter and energy could cascade outward in a web of interconnected effects that alter the shape of the community as a whole.

Many stories include some of these plot elements without ever approaching anything like ecological storytelling. So, for example, Charlotte Brontë's classic novel *Jane Eyre* (1847) includes surprise inheritances, a deadly epidemic, and a disastrous sexual relationship. All these occurrences play some role in Jane's development and her final fate. But in *Jane Eyre,* the significance of these events finally turns on how they affect Jane's psychology: her happiness and the happiness of her small circle of family and friends. Whatever information *Jane Eyre* contains about the relationship between nature and society in the 1800s is not central to the plot; it is passed along only in fragmented episodes and asides. It takes a canny critic to identify such moments, to pause over them, and to make arguments about why they might matter from an ecological perspective.[8] Brontë herself is not attempting a concerted investigation into how communal interdependence actually works, even if we can find lessons on that topic when we read her from our own more ecologically attuned perspective.

Ecological plots, by contrast, are interested in individual fates primarily for the ways in which they illustrate broader communal entanglements. Ecologically minded works narrate in order to show how a given character is intimately bound to much larger, often unacknowledged communities. These stories are bent on revealing how beings affect one another at a distance and across scales, how we participate in feedback loops and reciprocal interactions that tend to be hidden from our everyday experience. In essence, the ecological plot is an imaginative tool for revealing these dynamics, for making them visible and thinkable in a way that daily life does not. An unconstrained ecological plot overleaps conventional borders of social storytelling. It blows through the conceptual walls we erect on a daily basis between countries or classes or species, the fictional divisions we impose on the world to make it seem small enough and orderly enough to understand. Ecological plots follow the consequences of physical interactions wherever they may lead, revealing surprising and sometimes unnerving patterns of interdependence. They foreground those networks of material exchange between individuals, populations, and species that compose the "circuit called the biota."

The ecological plot helped give rise to ecology, but it became a popular narrative form in the English-speaking world long before ecology was recognized as a science. In fact, this mode of storytelling is striking for the way its history connects fields that now seem utterly unrelated to one another. As it circulated freely among a variety of thinkers in the early 1800s, the ecological plot played a formative role not only in the rise of ecology but in the rise of economics and of the realist social novel as well. It first became an object of public debate in the late eighteenth century, when ecological plots lay at the heart of bitter disputes over political economy—the field that would eventually give rise to modern economics. Next the ecological plot transformed the realm of literary fiction, as progressive novelists began to recognize that its depiction of interconnection could encourage an inclusive understanding of community among their more hidebound middle-class readers. The ecological plot only infiltrated natural history later, where it eventually spawned the new science of ecology and revolutionized the study of how living communities work.

The history elaborated here begins in the exhilarating exchanges among political economy, natural history, and realist fiction that were still opening up in the early nineteenth century. It concludes with the disciplinary ruptures between economics, ecology, and the social novel at the century's end. Mapping the trajectory of the ecological plot as it surfaced in a diverse set of writings over the course of the 1800s traces one illuminating path through these intellectual shifts. It helps clarify how a profound and widely shared understanding of material interdependence that emerged in the nineteenth century ended up producing three very different visions of collectivity in less than a hundred years.

These divergent approaches to community are more than just an academic concern. We live in a time when worldwide struggles for environmental justice are regularly brought to a standstill by a perceived incompatibility between the priorities of ecology and those of economics—and both disciplines can prescribe courses of action that contradict everyday standards of humane behavior. The escalating crises of habitat destruction, toxic pollution, mass extinction, and climate change have made it clear that environmental sustainability conflicts with the routine assumptions of life in an industrialized, profit-driven society premised on growth. Citizens around the globe are now engaged in urgent conversations across regional, national, and disciplinary boundaries, trying to reconcile the clashing visions of modern economics and modern ecology. What they are searching for is some way to harmonize these competing worldviews with the ideal of a flourishing, humane community. This book serves as a reminder that the divergent approaches to interconnection typical of the modern West in fact possess a common inheritance. A single, shared form of storytelling lies at their root: the ecological plot. One way to begin bridging the gaps between our conflicting visions of communal prosperity is to learn more about their shared origins—including the formal and philosophical disputes that slowly drove them apart.

At the very least, then, this literary approach to ecology aims to deliver a fresh perspective on the history of ecological ideas. But I hope its interdisciplinary genealogy will do more, inspiring not just a fresh way of thinking about the past but also fresh ways of thinking about the future. In laying out how influential figures from a variety of fields have used writing to imagine connections across species over the past two centuries, this book reminds us of the many kinds of interdependence, intellectual as well as material, that sustain us. These forms of interdependence must be respected if we are to weather our mounting crises and emerge both materially and morally intact.

ECOLOGICAL IDEAS, LITERARY TECHNIQUES

While the literary history presented here is new, this is hardly the first book to insist on a relationship between literature and ecology. By now, scholarly attempts to understand what fiction, poetry, drama, and nature writing can tell us about the environment are so plentiful that this area of expertise has earned its own name among literary critics: it is called ecocriticism. But most studies of the intersection of literature and the environment proceed by looking through literature for examples of content or themes we now consider environmental. So, for example, some ecocritics have located the roots of modern environmentalism in the religious respect for nature encouraged by Romantic poetry. Others have found it in the love of wilderness spread through the writings of transcendentalists like Ralph Waldo Emerson and Henry David Thoreau. Still other ecocritics have searched through past literature for evidence of environmental anxiety or emerging signs of nature in crisis. This thematic approach has produced valuable insights into how early modern poets understood fossil fuels, for example, or how Victorians responded to dwindling natural resources and the thick carpet of smog that hung over industrial London. Whatever the method, all forms of ecocriticism work to piece together a more complete cultural history of attitudes toward the environment, one that includes the growing public awareness of threats to the natural world.[9]

This book is different. Instead of searching literature for concepts related to nature and its exploitation, it aims at something more basic: a history of the forms of writing that made ecology thinkable in the first place. I track a single kind of storytelling, the ecological plot, as it was taken up and altered by various thinkers over the course of the nineteenth century. This emphasis on a single literary form sets this study apart, distinguishing it not only from typical ecocriticism but also from histories of science. Historians generally search for ecology's antecedents in emerging scientific theories or policies toward natural resource management.[10] Yet the wide dispersal of ecological storytelling in the nineteenth century presents a challenge to such orderly historical approaches; the writers who adopted such storytelling often defy disciplinary classification.

The literary history explored here, for example, begins and ends naturally enough with chapters that consider the work of Charles Darwin. Aside from these Darwinian bookends, however, I take what may look like a circuitous route through nineteenth-century writing. To start, I locate the cornerstone of modern ecological, economic, and literary storytelling in the widely despised arguments of a single man: the clergyman and demographer Thomas Robert Malthus. I next examine the once beloved but now almost forgotten novellas of Harriet Martineau, an economic popularizer who served as a sort of linchpin holding fiction, natural history, and political economy together. Later I use urban fiction by Elizabeth Gaskell, novels and essays by George Eliot, and fiction by Thomas Hardy as waypoints that mark growing rifts between political economy, natural history, and the social novel. These are the rifts, I suggest, that would eventually produce the clashing prescriptions from economics, ecology, and humane moralists that make environmental problems so difficult to solve to this day.

In short, this study does not linger over the kinds of nature writing or environmental problems we typically associate with ecology. But even if the writers and issues explored here have rarely been linked together in this way, combining them in a single history makes sense. After all, the intellectual traffic between Darwinian natural history and nineteenth-century economic thought is well known. The terms *ecology* and *economics* are even etymologically linked; they both begin with *eco-*, an English rendering of *oikos*, the Greek word meaning "household" or "home." The close association between economics and the life sciences has long been a point of pride for advocates of social Darwinism, for supporters of sociobiology, and for political leaders who champion the free market as the most natural force for socioeconomic development. At the same time, protests against the domineering logic of free-market economics lay at the root of much Romantic and Victorian literature, as cultural historians have long noted. And plenty of studies have argued for the close connections between fiction writing and natural history, both before and after the Darwinian revolution.[11]

There is nothing new, then, about establishing links between political economy and natural history, or between political economy and fiction, or between fiction and natural history. Each individual side of this triangle is well documented. What is missing, however, is a unifying account of how these fields came together in the nineteenth century as well as how they slowly broke apart. This book is a preliminary attempt at that synthesis. It is a daunting task: the multitude of ideas and arguments exchanged among these fields (each of them sprawling and variegated in its own right) makes it easy to get lost in them. What renders this effort possible at all is a relatively single-minded literary focus, a commitment to tracking one very specific literary structure as it moved between fields, changing and developing in the hands of different practitioners. This specific literary structure is the mode of storytelling I am calling the ecological plot.

The phrase "ecological plot" is my own coinage, but the phenomenon it names should be recognizable to anyone familiar with novels written during the Victorian era. The narrators of books by Charles Dickens, William Thackeray, George Eliot, and others explicitly discuss the networks of connection that structure their plots and capture their notions of how society operates. Literary scholars have used any number of names to refer to these expansive, interconnected plot structures. The great critic Raymond Williams understood such novels as attempts to represent "knowable communities," to produce single books that each contained "a whole community, wholly knowable." It was a goal that felt at once more necessary and less attainable than ever in the Victorian era, a time when modern industrial routines began to interrupt the established relationships of an older social order. Marxist theorists have long observed the novel's drive toward community-mapping as an attempt to represent some kind of social totality, a coherent collective that is never experienced directly except in a series of seemingly disconnected and brief interpersonal encounters. Whatever their politics, scholars tend to agree that Victorian novels construct "networks that unfold temporally," as

the critic Caroline Levine puts it, producing what Gillian Beer has called the characteristic "multi-plot form of many of the greatest Victorian fictions." These elaborate, entangled plot structures work by "combining the amplitude and arboreal form of their large narratives with an increasing insistence on the moral duty to recall and reconnect." The gradually blooming networks of interconnection that Victorian fictions explore all depend on "the long narrative arc of the multiplot novel" that is so closely associated with this historical period that many critics, such as Tina Young Choi and Peter Garrett, explicitly call it the "Victorian multiplot."[12]

Redescribing this narrative form as ecological helps emphasize its significance to intellectual history, a significance that extends far beyond the pages of nineteenth-century fiction. The basic argument of this book is that ecological plots underlay key nineteenth-century contributions to political economy and natural history as well as realist novels. Nor was this shared plot structure a simple coincidence. It is possible to trace the ecological plot's emergence into public debate and its subsequent spread among important contributors to all three fields—contributors who would come to be considered the founders of the now discrete disciplinary practices of economics, ecology, and the social novel. Given the centrality of the same structure to all three fields, it may seem odd to give it a name that sounds exclusively ecological. But calling it the ecological plot clarifies that it is the structure that helped give rise to the science of ecology in particular. The fields or discourses of political economy and realist fiction both predate the rise of this kind of storytelling in the modern Western world, even though all these categories of writing would be transformed by its vision of interconnection. As thinkers in each field took up this form of storytelling, they reshaped it to their own ends, emphasizing different aspects of interconnection and valuing such connections differently. In the process, they altered both the formal features and the moral implications of such plotting. These modifications finally resulted in the seemingly incommensurable approaches to community that define these fields today.

This, then, is a story of rupture and discord. But it starts in pre-disciplinary synthesis, and it aims to inspire renewed faith in the possibilities that can emerge from such cross-disciplinary conversation. In this spirit of synthesis and conversation, I have tried at every turn to minimize jargon and to avoid lengthy discussions of the many scholars who have touched on portions of these topics before. No doubt there are upsides to the adoption of technical terms, just as there are upsides to arguments made through explicit debate with fellow specialists. Nevertheless, both these conventions diminish the accessibility of scholarly writing. So when a specialized term I use may be unfamiliar to the general reader, I try to provide a definition right in the body of the text when it's first introduced. When definitions or histories are the subject of some debate (as they often are in the humanities), I have tried to gesture at those debates in the place properly reserved for existing scholarship: the notes.

Another academic habit particular to literary studies is a tendency to assume that readers are familiar with both the books under discussion and the general historical developments surrounding them. This habit saves time and space in writing, but it alienates readers who have not devoted years of study to a vast body of literature, history, and criticism. So whenever I introduce historical figures, literary texts, or larger contexts here, I do my best to provide an overview of who and what I am discussing to help orient curious readers. Books are summarized in as much detail as necessary, but in as little detail as possible—whatever it takes to explain evidence and get the larger points across without giving too much away. (Sometimes, spoilers are inevitable.)

A final habit of literary critics that is alien to outsiders is the habit of drawing a sharp distinction between the author of a work and the narrator or speaking voice within that work. This distinction helps differentiate writers' stated intentions from the effects—intentional or not—of the work they produce. Separating authors from their works is especially crucial to discussions of complex, experimental genres such as the dramatic monologue, modernist and postmodern texts, and autofiction. But constantly drawing lines between narrators and writers is confusing and unnecessary for the purposes of this particular study, so I have mostly abandoned the habit for clarity's sake.

In the end, all writing is torn between two opposing poles: the ideal of comprehension and the ideal of comprehensiveness. A short, straightforward book increases readers' comprehension. It is easy to understand, but it risks reducing the overwhelming complexity of the world to a few oversimplified points. A longer, more convoluted book can tackle that complexity, but it risks burying significant ideas under the welter of detail necessary to demonstrate its comprehensiveness. Achieving a balance between these two ideals is—like all worthy efforts—probably impossible. Nevertheless, this book represents my best attempt to find some kind of middle ground. In it, I generally prioritize clarity and concision over exhaustive, encyclopedic coverage. Most of the points I make are exemplified through the work of a single author or, occasionally, a small handful of writers.

In streamlining this history, however, I have tried never to oversimplify it. The authors discussed here were chosen because they are representative of some larger movement or because they were hugely influential on subsequent thinkers. In most cases, both criteria apply. Regardless, the relatively narrow selection of writers and thinkers included in this book is intentional. Keeping this history confined to a small handful of memorable individuals makes its claims easier to understand and to follow over time. Much has been omitted that might have been included. There is little discussion, in particular, of intellectual or historical developments beyond the confines of Britain. That is partly because Britain really was the crucible for the ideas and fields discussed here, but it also reflects my efforts toward concision.

This focus on Britain has important geopolitical implications. Everything covered in this book unfolds during the ascension of the British empire, and imperial policy and plantation economics are entangled in the rise of the ecological plot in ways that will be apparent from some of the earliest texts I discuss. As the critic and theorist Bénédicte Boisseron points out, a failure to acknowledge race and empire when discussing questions of the nonhuman is dangerous. It reinforces a longstanding scholarly habit of consigning non-European places and nonwhite peoples to the category of "that which essentially does not concern us," a habit doubly pernicious in fields like ecology and animal studies, given that notions of race and species have been entangled from the start.[13] It would be irresponsible to maintain the silence around questions of race and empire that has plagued studies of the natural world for so long. It would also run counter to the purpose of this book, which is to trace how ecological storytelling usefully expanded understandings of ethical community during the form's rise in the nineteenth and twentieth centuries. So whenever ecological plots lead the figures treated here to recognize their own complicity in racist and imperialist systems—in whatever partial and inadequate ways they do so—I try to provide context on those systems and what role ecological plots played in making them visible. Yet the strategy of following this storytelling structure as it spread within a small group of influential thinkers has made it impractical to undertake anything close to a thorough treatment of the devastating impacts of racism, colonization, and empire in the nineteenth century. I try to note the shortcomings of the thinkers I cover on these issues, both because they have been damaging in their own right and because those of us steeped in modern Western ways of thinking have inherited their flaws and oversights. But the approach I have adopted here prevents me from giving such injustices the full scholarly treatment they deserve.

These are painful tradeoffs. Certainly much has been lost in depth, comprehensiveness, and political incisiveness that might have been included in a different kind of project. But hopefully those losses are offset by gains in the book's practical value as a roadmap for many different kinds of readers interested in exploring the writings that contributed to the modern developments of ecology, economics, and socially committed fiction. It would be a pleasure to see others fill in the gaps, extend the boundaries, or even redraw the contours of the sketch I attempt here. It is to that sketch in its broadest outline that I now briefly turn.

FROM SHARED STORIES TO CONFLICTING DISCIPLINES

Like many histories, this one starts with a fight.

In the late 1700s, a young clergyman living with his parents outside London found himself annoyed by the utopian social ideas in vogue across Europe. He began writing his objections down, and before long he accrued enough material to publish an extended tract on the topic. When it appeared, the *Essay on the Principle of Population* (1798) catapulted the Reverend Thomas Robert Malthus to sudden fame—and equally sudden infamy. After a few years of public furor, the work Malthus began in the *Essay* finally landed him a comfortable professorship in history and political economy. But the same writings inspired endless attacks from his Romantic contemporaries. Their target was Malthus's gloomy insistence on the dangers of overpopulation, starvation, disease, sin, and misery. The volleys of criticism from William Godwin, Percy Shelley, and many other major names made a lasting dent in Malthus's reputation; he remains a figure of distrust and outright scorn to this day. But the controversy that hung over Malthus also ensured that his name remained prominent and his *Essay* stayed in print, enabling it to find new readers with each coming generation. As it turned out, those readers would go on to shape a sizable portion of the arts and sciences we recognize today.

The first chapter of this book returns to Malthus's misunderstood *Essay* with unclouded eyes, paying especially close attention to the work's literary form. It argues that the anecdotes at the heart of Malthus's work share a peculiar and innovative approach to storytelling. In them Malthus traces how a single, apparently isolated decision has ripple effects as it spreads across a materially interconnected community—a community that is defined not simply by human stakeholders but also by the many plants and animals on which human beings depend. Malthus's *Essay* thus marks the first highly visible and influential appearance of ecological plotting in the modern world. While many despised Malthus, two of his acolytes would help spread this form of storytelling to the life sciences and to literary fiction. The first chapter of this book opens with Malthus, but it closes with Harriet Martineau and Charles Darwin. Both of these thinkers adopted and deployed Malthusian plot structures in the 1830s, using them as scaffolding for their pioneering work—for Martineau in Victorian social fiction, and for Darwin in the life sciences, including ecology. Together Malthus, Martineau, and Darwin created a way of telling stories without backgrounds, narratives that encouraged careful accounting of all the living and nonliving beings entangled in webs of material interdependence.

Yet this cross-disciplinary harmony would not last forever. The second chapter moves forward historically to the social turmoil of the 1840s, when the political economy so prized by Malthus and Martineau failed to deliver on its promise to maximize collective happiness—or even to provide the starving masses with enough food to survive. Delving into Elizabeth Gaskell's *Mary Barton* (1848) and other social-problem novels of the 1840s and '50s, the second chapter shows how Victorian novelists adopted the mantle of humane moralizers in the face of such suffering, a position that set them apart from political economists like Malthus and naturalists like Darwin. Nevertheless, Gaskell and others continued to rely on ecological plots borrowed from Malthus and Martineau, using these narratives of material interdependence to stress how moral duties to others extended across classes and brought communities together. This emphasis on the moral dimensions of interdependence constituted a first step toward future ruptures between these adjacent modes of writing about community. In the coming years, scientists and economists increasingly focused on the dynamics of material exchange without any consideration of their implications for ethics.

This growing rift between matter and morality found its most decisive expression in the midcentury writings of George Eliot. Before she wrote the fiction that made her famous, Eliot (who went by Mary Ann or Marian Evans) thoughtfully described the tensions among political economy, natural history, and realist fiction. Chapter 3 uses Eliot's career arc to explore the break between these fields. The chapter opens with "The Natural History of German Life" (1856), Eliot's review essay on the fraying relationship between these three attempts to understand community. It then charts how Eliot's early admiration for natural history dwindled as she realized the novel's unique ability to emphasize the flow of moral energies over and above material ones. While material interconnection structured Eliot's storytelling throughout her career, the shift from her first novel, *Adam Bede* (1859), to her second, *The Mill on the Floss* (1860), demonstrates an increasingly single-minded commitment to focusing stories on humane feeling. As Eliot saw it, this moral focus required downplaying the close relations of humans and nonhumans that first drew her attention in works of natural history. If Malthus, Martineau, and Darwin had begun to develop an ecological way of telling stories without backgrounds, Eliot applied her artistic and intellectual talents to the task of restoring such backgrounds, excising the natural world from a realist tradition that had only just begun to include it. Her later novels rely on the quiet power of literary form to reassert the distinction between character and setting, helping enforce the priority of humans by obscuring our connections to the environments that sustain us.

The fourth and final chapter drives home the outsized effects these seemingly minor formal distinctions can have on our ability to see the world ecologically. This chapter focuses on the late Victorian writings of Thomas Hardy, a novelist often admired for his ecological intuitions about humanity's embeddedness in the natural world. Yet putting Hardy's descriptions of heathland settings in *The Return of the Native* (1878) side by side with Darwin's accounts of the same places in *On the Origin of Species* (1859) exposes very fundamental oversights in Hardy's breathtaking natural descriptions. These oversights led Hardy to misunderstand not only the natural landscape but also—crucially—humanity's role in maintaining it. What Darwin saw (and what Hardy could not see) was that heathlands were fragile, man-made habitats, uniquely fleeting ecosystems that could only be stabilized by ongoing human customs of grazing and cultivation. Hardy's mistaken interpretation of heathlands as stubborn, primordial, inhuman places inadvertently fueled the negative attitudes toward these habitats that led to their near-total eradication over the first half of the twentieth century.

The contrasting accounts of English heathland in Hardy and Darwin underscore just how far realist fiction had diverged from ecological storytelling by the final decades of the nineteenth century. They suggest, too, how difficult it was—and is—to see ecological interdependence through the realist novel's highly moralized, anthropocentric lens. But the point of this history is not to decry such oversights or to lament the gaps that eventually separated the once intimate fields of political economy, natural history, and social fiction. There were benefits to the increasing specialization of each of these fields as they evolved during the 1800s. Economics grew in precision and explanatory power as it abandoned its origins in moral and political philosophy. It assumed new authority as a more abstract, more mathematical field. For their part, the life sciences shook off the theological shackles and taxonomic quibbles characteristic of so much natural history. By the dawn of the twentieth century, biologists and ecologists were formulating increasingly objective, empirically testable accounts of how interactions between organisms determined the shape of communities. And as the social novel gained independence from the doctrines prescribed by these adjacent emerging disciplines, it assumed its now almost exclusive focus on human life. In the process the novel acquired a newfound cultural authority both as an art form and as a source of insight into the human condition.

The benefits of all this specialization came at a price, however. True, economists developed neatly precise models that explained how prices of goods and labor fluctuate in a global marketplace. At the same time, they bought into and broadcast the fallacy that markets are logical, inevitable systems that accurately indicate the value of goods through pricing alone. The moral acuity of novelists made them vigilant against the selfishness and the reductive valuations that underlay economic thought. Yet their social novels achieved moral clarity only by overlooking the complexities introduced by the nonhuman contributors to social life. The modern life sciences did a far more thorough job of tracing the material dynamics that bound nature and society together. But they achieved material insight only by deadening themselves to the moral questions before them, questions ranging from the pain inflicted on laboratory animals to the ethical duties implied by interconnection.

In its own way, then, each approach to interconnection gained rigor at the expense of amplitude. With expertise came a kind of narrowness that left certain material and moral conflicts not just unresolved but even unacknowledged. We are living with the consequences of those unacknowledged conflicts today. They surface repeatedly in every fight over the environmental devastation of economic development, or the unbearable costs of sustainable energy transitions, or the injustices that arise from unchecked capitalism (on the one hand) and mainstream environmentalism (on the other). The conclusion of this book briefly surveys a few of the attempts over the last century to bring these fields back into conversation with one another, touching on a sampling of landmark literary experiments to imagine a better world.

Obviously, the brief history sketched out in this book will not solve these problems. In showing how these visions of community departed from one another, however, it may help us come to grips with the narrowness and omissions of each approach. And in reminding us of their shared inheritance, it may suggest new ways to tell the story of how we got here and where we are going, alternative narratives that could help us overcome such narrowness and open up possibilities for a more coherent, more sustainable, and more equitable future.

CHAPTER 1

Approaching Ecology

When Susan Darwin sat down on October 15, 1833, to write a letter to her brother Charles, their surroundings could not have been more different. Susan was living comfortably at The Mount, a Georgian estate built by their father outside the English market town of Shrewsbury. She, Charles, and three of their siblings had been born at the house, and Susan would reside there for the entirety of her life. She was, in short, at home in every sense of the phrase, visiting with established family friends and imagining the same sort of future for her younger brother. "I quite long for you to be settled in just the same kind of manner my dear Charley," she confided to him.[1]

At the moment, though, Charles was thoroughly at sea. After graduating from Cambridge in 1831, the directionless younger son of Robert and Susannah Darwin had accepted a spur-of-the-moment offer to travel the world as the resident naturalist aboard HMS *Beagle,* a surveying ship circumnavigating the globe. In October 1833, the *Beagle* was anchored near the mouth of the Río de la Plata, mapping the intricacies of the shoreline between the tumultuous Argentine Confederation and the newly independent Uruguay. Susan's autumn afternoon was thus a spring morning for her brother, who had gone ashore several weeks earlier on a journey inland to explore the Paraná River. Despite suffering from poor health, Charles found time to marvel at everything from the lopsided, scissorlike beak of the black skimmer (*Rynchops niger*) to the river's impenetrable clouds of mosquitoes. "I exposed my hand for five minutes," he observed one night, "and it was soon black with them; I do not suppose there could have been less than fifty, all busy sucking."[2]

Nevertheless, despite the great gulfs separating her from Charles, Susan found a rather unexpected source of connection between them: her recent leisure reading. Charles was already sending home his elaborate notes on the animals, plants, and landforms he encountered. Susan read his letters and journal entries alongside a set of fictional tales that paid similarly detailed attention to the natural world. Noticing the parallels between the two bodies of writing, she mentioned the luxuriant fictional landscapes to her brother. "I have just been reading an account of Ceylon in a kind of novel called 'Cinnamon & Pearls,'" she told him. "[T]he descriptions of the vegetation are so beautiful that I don't wonder you have a great desire to go there."[3]

Although she could not have known it at the time, Susan's unassuming note marked the start of an intellectual conversation that would revolutionize modern science. It was the first in a series of exchanges that brought three major nineteenth-century thinkers into mutual conversation: Thomas Robert Malthus, Harriet Martineau, and Charles Darwin. Taken together, the interwoven ideas of these three writers proved foundational to the arts and sciences as we understand them today, shaping the practices of economics, ecology, and—in what might sound like an outlier in this list—the realist novel. These fields now seem easily distinguishable from one another. But that was not always the case.

Economics, ecology, and the realist novel share a complex history rooted in a shared commitment to the dynamics of living communities. Modern understandings of such dynamics developed rapidly over the course of the nineteenth century thanks to the intellectual constellation formed by Malthus, Martineau, and Darwin. In different ways, each of these writers helped the Western world wake up to a fact long established among Indigenous cultures around the globe: namely, that human societies cannot be isolated, either logically or materially, from the natural world on which they depend.[4] The dissemination of this idea eventually gave rise to the environmental movement and other modern efforts to think conscientiously about how humans should behave toward the natural world.

In the England of the 1830s, however, serious reflection on the material interdependence of humans, plants, and animals was rare. It was largely confined to the field of political economy—a precursor to modern economics that tried to explain why some countries ended up wealthier than others. Political economists had linked the possession of natural resources to national wealth, and their theories were gaining traction among academics and politicians seeking to understand how society worked. Nevertheless, political economy was a notoriously dry and jargon-ridden subject, one inaccessible to all but the most committed scholars. The general public was unable to focus on the arid logic that was coming to define the field. Many people remained skeptical of the dictates of political economists, these newfangled experts who doled out advice and crafted policies that flew in the face of both tradition and common sense.

"Cinnamon and Pearls" (1833), the story Susan Darwin mentioned to her brother, was part of one woman's extraordinary effort to warm a reluctant populace to the new social science. It was the eighteenth tale in Harriet Martineau's *Illustrations of Political Economy* (1832–34), a set of short fictions Martineau produced to make political economy intriguing to a mass readership. Unpromising as that mission may sound, the novellas became bestsellers, launching the unknown Martineau to stardom almost overnight. Written and released serially over the course of two years, the *Illustrations* introduced her into all the prominent literary and political circles of the era. These circles included the social set of the Darwins themselves, where Martineau's unflappable analytic mind made her fast friends with Charles's older brother, Erasmus Alvey Darwin.

From the perspective of literary and intellectual history, there are two things that made Martineau's work vital to the shape of the world to come. The first was her peculiar mode of storytelling—a form of narrative I call *the ecological plot*. Like the ecological science that would emerge from Darwin's writings in the second half of the century, Martineau's *Illustrations* used narrative to trace the exchanges of matter and energy that knit individual beings together into complex webs of interdependence. This formal feature was likely what drew Susan's attention to the lush plant life of Sri Lanka in "Cinnamon and Pearls." Still, any similarity between Martineau's plot structures and Darwin's later writings could be written off as mere coincidence. It is the second noteworthy feature of Martineau's work—the conditions of its production and dissemination—that makes such a coincidence highly improbable. Almost immediately after Martineau's first novella was published, she developed an unusually wide readership and an extensive social circle to go along with it. Taken together, these two features of Martineau's work made her the linchpin of an exceptionally significant network of thinkers in early nineteenth-century Britain. That network eventually helped spur the rise of modern economics, ecology, and realist fiction.

This chapter traces that surprising story. It begins with the dim recognition of interdependence that arose in the late 1700s in the influential but widely derided writings of Thomas Robert Malthus. Malthus's astonishing *Essay on the Principle of Population* (1798) laid the groundwork that has supported economics, demography, and ecology ever since. Despite having passed her own childhood steeped in the many popular objections to the *Essay*, Martineau grew to admire Malthus—first his work, and then the man himself in her brief friendship with him in the final years of his life. I then dive into the ways that Martineau expanded Malthusian ideas of interdependence into the elaborate plot structures that underlay her hugely successful fictions. The chapter concludes with an examination of the joint influence of Malthus and Martineau on Darwin, who drew on their insights in his groundbreaking treatise *On the Origin of Species* (1859). Darwin's book established the future direction of the life sciences by laying out both the theory of natural selection and the understanding of interspecies communities that gave birth to ecology. Tracing this brilliant constellation of protodisciplinary thinking in the first half of the nineteenth century clarifies how the act of imagining material exchange helped found multiple modern fields premised on the interdependence of living things. It also sets the stage for the later history of how these modes of making sense of community would eventually break apart.

MALTHUS AND THE ECONOMY OF INTERCONNECTION

Daniel Malthus was getting on his son's nerves. The young Thomas Robert—friends called him Robert or simply Bob—had just graduated from Cambridge. In order to save money, Robert had moved back to the family home in Surrey, conveniently located within commuting distance of his new job as curate at a local church. For the first few years of his professional life, Robert lived and dined primarily under his parents' roof. There he read news of the French Revolution and listened with growing impatience as his progressive father extolled the virtues of Jean-Jacques Rousseau, William Godwin, and other radical thinkers of the age. After close to a decade of this talk, Robert had had enough.[5]

In a reversal of the stereotypical relationship between youthful idealism and cynical experience, Robert was sure that his father's utopian dreams had no underlying substance. The way Robert saw it, Daniel's idols—people like Godwin and the Marquis de Condorcet—failed to ground their radical ideas of human perfection in reality. Following a particularly spirited exchange over dinner one night, Robert decided it might be easier if he wrote out what he was trying to say. According to his own account, he "sat down with an intention of merely stating his thoughts to his friend"—that is, his father—"upon paper, in a clearer manner than he thought he could do in conversation."⁶ Before long, the young Malthus's explanation grew to the length of a short book, and he decided to see if any publishers would find it worthy of wider distribution. As it turned out, one did.

An Essay on the Principle of Population, as it Affects the Future Improvement of Society, with Remarks on the Speculations of Mr Godwin, M. Condorcet, and Other Writers has been at the center of controversy since it first appeared anonymously in 1798. As the title suggests, its argument hinges on the relationship between prosperity and population, with special attention to the suffering that comes about when a population outstrips its own food supply. When the authorship of the *Essay* became known, it jump-started Malthus's career as a political economist. It also cemented his reputation in many circles as a moral monster. Angry readers cast Malthus as a dour clergyman who predicted that the human species was doomed, a hypocrite who blamed the poor for breeding but had no qualms about siring several children himself. Even present-day economists who admit Malthus's significance to the field are quick to dismiss the *Essay on the Principle of Population* for its bad math and its failure to account for technological innovation. By 1805 the work was so notorious that *Malthusian* had entered the English lexicon, where it is still invoked as a slur rather than a badge of honor. Malthus experienced a brief renaissance in the 1970s and '80s among environmentalists concerned with resource exhaustion, only to fall out of favor again when critics observed how overpopulation was being weaponized as an ideology that enforced racist and classist social policies.⁷

Yet people who dismiss Malthus seem either not to have read his work or not to have taken it seriously. What they attack is a lazy caricature of the man and his ideas, a caricature first sketched by the Romantics and utopians most threatened by the pragmatic bent of the *Essay*. Returning to the first edition of Malthus's *Essay* reveals a surprisingly evenhanded and even progressive document—one whose argument against radical social thought has nothing to do with future fears of overpopulation. If Malthus seemed like a drag to utopian thinkers, it was because he brought their airy social fantasies back to earth in a very literal manner. Re-rooting society in the soil, Malthus insisted that human communities would always be shaped and constrained by the earth, including the plants and animals that share it with us. The matter that goes into making human bodies, he pointed out, can all be traced through such chains of nonhuman organisms. His careful explication of this basic fact upset a line of revolutionary thinkers who liked to fantasize that growth and progress could continue without limit forever. But if Malthus's writings offended some revolutionary thinkers, they also planted the seeds for other forms of revolutionary thought that would reshape literature and science in the decades to come. Both because it proved so valuable and because it has been so widely misunderstood, Malthus's short book is worth examining at length.

The key insight of Malthus's "sketch of the state of society" is that the growth of human communities depends on the nonhuman communities that sustain them.[8] This idea was not entirely new among political economists. An earlier group of French thinkers, the Physiocrats, had placed a similar emphasis on food and agriculture as the origin of all wealth. That idea led the Physiocrats to develop a marked suspicion of money itself, which they saw as a dangerous and mystifying abstraction. Currency, they pointed out, made it possible to assign nominal values to goods and services that were actually useless to sustaining human life.[9] From the Physiocrats, then, Malthus inherited a conviction that the only sound basis of value lay in material goods capable of producing and maintaining life—what he called "real funds for the maintenance of labour," funds that could be either literal food or any kind of currency that was directly "convertible into a proportional quantity of provisions" (125).[10]

What set Malthus apart from the Physiocrats was his realization that such material funds were limited. As the historian Fredrik Albritton Jonsson notes, other Enlightenment theorists tended toward utopian flights of fancy; they persistently slipped into romantic ruminations on "the benign system of nature" and the future potential of human society.[11] Malthus, by contrast, "diverged from [both Adam Smith and the Physiocrats] in stressing the feebleness of technology and knowledge in the face of geographic restraints and demographic forces."[12] The restraints Malthus identified were not forces in some invisible, abstract sense of the term. They were concrete, physical limits to growth—limits defined by the material realities of plant and animal bodies.

Throughout the *Essay*, Malthus undercuts abstract musings on social progress from Godwin, Condorcet, and others by returning to the concreteness of the soil, vegetation, and animal flesh required for society to function. A central assumption of Enlightenment utopianism was that human science and technology had improved dramatically and promised to do so indefinitely. (Nearly identical assertions remain popular among economists who dismiss Malthus today.) Malthus refutes these assumptions early in the first edition of the *Essay* by turning attention away from humans entirely, focusing instead on wheat, cabbages, carnations, and sheep. By selective breeding, he notes, the features of these living beings are subject to inspiring improvements, just like other human technologies. By the late 1700s, breeders were so successful at developing new kinds of livestock that they believed it would be possible, given enough time, to alter and improve a breed to promote any features a breeder might desire. "[I]t is a maxim among the improvers of cattle that you may breed to any degree of nicety you please," Malthus notes, because "some of the offspring will possess the desirable qualities of the parents in a greater degree" (70).

Yet it would be a huge mistake to assume that continued success in altering breeds meant all creatures could be reshaped in any way imaginable. There are always limits to the plasticity of living bodies, even if breeders do not acknowledge them. "In the famous Leicestershire breed of sheep," Malthus points out, "the object is to procure them with small heads and small legs. Proceeding upon these [utopian] breeding maxims, it is evident that we might go on till the heads and legs were evanescent quantities." If the utopians were correct, careful breeding should be able to produce a new type of Leicestershire sheep reduced to headless and legless bundles of fluff lolling about in fields as they awaited harvesting for their wool and meat. "[B]ut this is so palpable an absurdity," he continues, "that we may be quite sure that the premises are not just, and that there really is a limit, though we cannot see it, or say exactly where it is" (70). Taking the Enlightenment dream of infinite progress to its logical conclusion, Malthus dreams up a laughable nonsense world where the average carnation is "the size of a large cabbage," a fantasy realm complete with swollen sheaves of wheat nodding beneath oaks as tall as mountains and spheres of headless, legless sheep wobbling contentedly among them (71).

Malthus's point is abstract and philosophical. He is critiquing a logical fallacy typical of his utopian opponents: the mistake of assuming that a failure to discover something (here, the limit of progress) provides persuasive evidence that the undiscovered thing must not exist. But Malthus makes his argument through metaphor, and the particular imagery he chooses does important work in proving his point. Focusing on sheep, flowers, grain, and trees forces readers to reconsider the logic behind claims of boundless human potential, but it allows those readers to do so without reference to the human at all. When we are asked to imagine limitless progress in the transformation of animal and vegetable bodies, the claim appears in all its naked absurdity, shorn of the self-congratulation that flatters humans' anthropocentrism—our implicit sense of our own centrality and excellence. Considered apart from human self-regard, it becomes clear that every change, progress included, has limits, even if we do not yet know what they are.

Treating the transformation of society in terms of the transformation of wheat and sheep also reinforces Malthus's larger point that these transformations are not actually separate issues at all. A human society's capacity for growth and development ultimately rests upon the growth and development of the plants and animals supporting it. The crux of Malthus's argument is precisely this inseparability, which he introduces through the devastating material limits of "room and food" (49). Food, he reminds his readers, is really just another name for a small set of edible plants and a handful of animals that eat them. Those plants, in turn, require a certain amount of soil and light to exist—finite resources that impose material limits on growth. Because humans depend on meat and vegetables, both of which finally derive from plants, which in turn derive from soil and water and sunlight, the whole of human society depends on the apportioning of soil, water, and sunlight among plants. Our material entanglements ensure that the fates of humans and nonhumans are never fully extricable from each other.

The "thought experiments and imaginative journeys" that make up so much of Malthus's argument are exercises in clarifying these entanglements.[13] They reconnect superficial indicators of wealth, prosperity, and progress to the physical resources that actually sustain human life. In the process, they drive home the ways economic quantification fails to capture the actual flows and obstructions of vital resources that move ceaselessly through a community that crosses conventional lines of class and species.

To see this reasoning at work, it helps to turn to a passage in the *Essay* where Malthus tries to imagine how higher wages might affect the neediest members of the working classes. Naturally enough, he begins by acknowledging the increased purchasing power enabled by pay increases. "Suppose," he writes, that "the eighteen pence a day which men earn now was made up five shillings; it might be imagined, perhaps, that they would then be able to live comfortably, and have a piece of meat every day for their dinners" (36). But then Malthus traces how that increased purchasing power would affect the prices of various food items: "The transfer . . . would not increase the quantity of meat in the country. There is not at present enough for all to have a decent share. . . . The competition among the buyers in the market of meat would rapidly raise the price . . . and the commodity would not be divided among many more than it is at present" (36–37). Next Malthus considers how this ripple effect could extend to reconfigure not just the price of meat but actual land usage across England. "If we can suppose the competition among buyers of meat to continue long enough for a greater number of cattle to be reared annually, this could only be done at the expence of the corn," he observes, "which would be a very disadvantageous exchange; for it is well known that the country could not then support the same population; and when subsistence is scarce in proportion to the number of people, it is of little consequence whether the lowest members of society possess eighteen pence or five shillings. They must at all events be reduced to live upon the hardest fare, and in the smallest quantity" (37).

The seemingly simple issue of wage increases thus leads Malthus to consider the amount of meat available in the country, the effect of the shifting demand on prices, the effect of those prices on farmers, the effect of those farmers on the relative populations of cattle and crops in Britain, and the effect of those altered cattle and crop populations on the welfare of the workers whose wages were increased in the first place. In this extraordinary extended reflection on the relationship between wages, cows, land, workers, and wheat, Malthus's critics see only one thing: Malthus seems to be arguing against wage increases. They conclude he must be a hard-hearted apologist for the status quo, a man who used environmentally attuned arguments to defend oppression and resist reform—and one whose clerical training enabled him to justify this hardness as the dispensation of divine providential law.[14]

The reality is more complex. In the first edition of the *Essay*, Malthus openly decries the extreme inequality of British society "as an evil," adding that "every institution that promotes it is essentially bad and impolitic."[15] While he assumed some social inequality was inevitable, he hoped that a more clear-sighted science of society would drastically reduce it. In the years to come, Malthus's transformation from obscure essayist to decorated political economist involved a massive expansion and revision of the *Essay*. After its first publication he traveled widely, gathered data on populations across Europe, and produced a second edition that more closely conformed to the expectations and practices of contemporary political economy. Among its new inclusions were far more extensive policy recommendations and more certainty about the lawlike increase of misery in tandem with an expanding population.

Yet the passages cited here, including the critique of inequality, remained largely intact in these later editions, albeit with assorted expansions and changes of wording. Malthus's desire to lessen human misery remains legible in those editions, too. But few readers have cared to look for it. The later editions' growing accord with mainstream political economy and their influential policy prescriptions for restructuring England's welfare system have eclipsed the first edition's vivacity, originality, and compassion in the popular mind. The result is a very partial recollection of Malthus's aims and impacts. The most important upshot of the *Essay* was not some defense of a cruel status quo; rather, it was an insistence on the need to think through the effects of apparently progressive policies as they rippled through the vast network of institutions and resources on which society depends. In every case, Malthus concluded his analysis of such ripple effects with an emphasis on how they would alter the distribution of life-sustaining energies across human, plant, and animal bodies.

This vision of a society supported through interspecies energy exchange marked a new chapter in British social thought. As the literary scholar Catherine Gallagher has argued, the *Essay*'s vision of energy transfer "radically reconceptualized the social organism," elaborating a new form of collective imagination that treated society "less as a metaphorical individual, such as 'the nation' or 'the population,' than what we now call an 'ecosystem.'"[16] Still, the language of organic wholes and the idea of "ecosystem" are too definite, closed, and comprehensive to apply to the worldview sketched out in the *Essay*. Malthus is typically depicted as a theorist of competition in an insular, claustrophobic environment. But his radical insight hinged not on closure but on openness—on a reimagination of community as a network of material exchanges across classes and species, exchanges that remained mostly unmapped. What Malthus made available, in short, was not the holistic idea of a closed and well-balanced ecosystem. It was an early Western version of what philosopher and critic Timothy Morton has called "the ecological thought," the open-ended recognition "that everything is interconnected" and that conventional distinctions between nature and society divide our entangled world in unhelpful and inaccurate ways.[17]

It is unfortunate, then, that Malthus is overwhelmingly associated with poverty and gloom rather than with blissful attunement to ecological interconnection. Yet in another sense that's appropriate, because Malthus calls attention to forms of entanglement that are not always uplifting. They are humbling in the original, etymological sense—they bring us low, returning us to earth with a rather rocky landing. As the critic Stacy Alaimo has warned elsewhere, "Attention to the material transit across bodies and environments may render it more difficult to seek refuge within fantasies of transcendence or imperviousness."[18] Actually tracing such transit requires acknowledging the surprising, sometimes unflattering ways that classes and species interact with and depend on one another.

For Malthus the project of tracing interconnection was essentially a work of realism—a written revision of previous theories about how the world worked, a corrective that used the details of everyday experience to provide a more accurate representation of life and society than prior, more idealized accounts.[19] While he admired the "enchanting picture" of human possibility that utopian thinkers painted "in the most captivating colours" (10–11), in the *Essay* Malthus promised to sketch a more realistic depiction of society "assisted . . . by actual experience, by facts that come within the scope of every man's observation" (28). A central element of Malthusian realism that distinguished it from its predecessors was its unflinching attention to the poor. "The sons and daughters of peasants," Malthus writes, "will not be found such rosy cherubs in real life as they are described to be in romances" (35–36). Here what may sound like an attack on the poor was actually an attack on the idealistic writers and thinkers who refused to see the poor for what they were: real people suffering terribly from unjust social arrangements.

Because romanticized depictions of poverty did not accurately describe the matter at hand, Malthus argued, they made it impossible to see such arrangements clearly—a necessary first step to correcting them. In fact, Malthus insisted, the problem of overpopulation followed by scarcity and starvation was not some hypothetical future issue utopian thinkers could solve at a later date. It was an ongoing problem with a long history. It had been ignored only because elite writers on social issues neglected to extend their chronicles of human communities to the lowest ranks of society. "[T]he histories of mankind that we possess are histories only of the higher classes" (20), he writes, so privileged and educated social thinkers cannot see "the perpetual struggle for room and food" (26) that has characterized society from its beginnings through to the superficially more stable, more harmonious present.

A well-known passage from the first edition of the *Essay* that discusses livestock and horseback riding illustrates the radical, transformative potential of this early ecological realism. It shows how connecting classes and species together in webs of interdependence changes our understanding of communal politics and ethics. Building on his earlier consideration of workers' wages and the rearing of meat, Malthus speculates that two of England's national passions—a love of beef and a love of horseback riding—could come together to starve and oppress the nation's poor. Because of the way matter and markets entangle, the seemingly separate consumer demands for meat and horses combined to discourage farmers from raising enough cereals to feed the British populace. "When the price of butcher's meat was very low," Malthus observes,

> cattle were reared chiefly upon waste lands [that could not grow reliable crops] . . . but the present price will not only pay for fatting cattle on the very best land, but will even allow of the rearing many on land that would bear good crops of corn. . . . I cannot help thinking that the present great demand for butcher's meat of the best quality, and the quantity of good land that is in consequence annually employed to produce it, together with the great number of horses at present kept for pleasure, are the chief causes that have prevented the quantity of human food in the country from keeping pace with the generally increased fertility of the soil; and a change of custom in these respects would, I have little doubt, have a very sensible effect on the quantity of subsistence in the country, and consequently on its population. (129–30)

Such passages suggest that Malthus's significance extends far beyond the tired, inaccurate stereotype of Malthus as a sullen prophet of overpopulation. He must be understood as an important figure not only to the foundations of political economy but also to the foundations of ecology, including its more overtly political and ethical forms.[20] In the sociologist Bruno Latour's words, today's political ecology is best understood as a technique for remapping community relationships, one that *"suspends* our certainties concerning the sovereign good of humans and things, ends and means. . . . This is its great virtue. It *does not know* what does or does not constitute a system. It does not know what is connected to what"—although it aims, through careful analysis, to find out.[21] Latour hints at an underlying kinship between economic analysis and ecological thinking, despite the differences between the fields. "Through the circulation of its tracers," including money and goods, Latour writes, economics "makes the collective *describable.*"[22] By following the circulation of food resources through some of the many humans, plants, and animals involved in the modern market system, Malthus inadvertently stumbled on the ecological reality of interdependence across species. He began developing a strategy of social analysis that expanded beyond the social to describe something like an interspecies collective.

Put another way, the *Essay* made ecology thinkable. One hundred and fifty years later, the writer and game manager Aldo Leopold—often credited as the founder of environmental ethics—would use the same ecological reasoning to urge a reconsideration of the selfish, anthropocentric behaviors that defined so much of Western modernity. "[E]nergy," Leopold explained, "flows through a circuit called the biota. . . . Land, then, is not merely soil; it is a fountain of energy flowing through a circuit of soils, plants, and animals. Food chains are the living channels which conduct energy upward; death and decay return it to the soil. The circuit is not closed; . . . but it is a sustained circuit, like a slowly augmented revolving fund of life."[23] For Leopold, recognizing the interconnectedness of this circuit or biotic community required a new ethic of interdependence, a land ethic that "enlarges the boundaries of the community to include soils, waters, plants, and animals, or collectively: the land."[24] Unfortunately, Leopold died fighting a brushfire before his most popular writings were published, so he never had a chance to witness their impact. They went on to provide the intellectual and moral grounds for the environmental movement of the second half of the twentieth century.[25]

Malthus's work appeared long before Leopold's, but his writing had no obvious impact on humanity's treatment of other species. In fact, while the ethics of the *Essay* were more progressive and complex than its critics admit, they remained unshakably centered on the human. That's because in addition to being a political economist, Malthus was an ordained clergyman who believed humans were the special focus of God's creation. The personal quirks of his theology sometimes upset the conservative Anglican establishment. Still, in the *Essay,* he never wavered in his underlying commitment to the sanctity of the human soul and "the importance of improving the happiness of the human species" (113).[26] Nevertheless, his redescription of society as part of an interspecies network of material flows had profound implications for the stories human beings told about the world and about our place within it. Those implications revolutionized the scientific community in 1859, when Darwin finally unveiled his Malthusian revision of natural history in *On the Origin of Species.* But Malthusian thought entered the literary world even earlier. There, too, it fomented its own quiet revolution. It helped extend the sense of who participated in the stories of living communities, destabilizing traditional notions of society by reminding readers of the complex, often inconvenient intimacies between humanity and the natural world.

HARRIET MARTINEAU'S MULTISPECIES STORYTELLING

Harriet Martineau did not fit the profile of a person likely to champion Malthusian thinking. As the educated daughter of a successful textile manufacturer in the early decades of the nineteenth century, she was expected to marry well and raise a family of her own. If she was unlucky on the marriage market, she could at least find work as a governess or schoolmistress, supporting herself respectably by teaching others the womanly virtues of middle-class domesticity. Almost any fate would have been preferable to becoming the public face of Malthusianism, a school of thought widely associated with misery, death, and the suppression of feminine nurturing instincts. Martineau herself absorbed exactly this sort of anti-Malthusian sentiment growing up. "I was sick of his name before I was fifteen," she recalled in her *Autobiography* (1877).[27] Yet a dramatic cascade of events in 1826 conspired to make Harriet Martineau the most influential Malthusian of her era. Her devotion to Malthus had, in turn, a transformative effect on both literature and science, giving rise to new ways of using narrative to understand the environment and our place within it.

Martineau's comfortable, conventional life began to unravel in 1825, when a financial panic caused twelve banks across England to collapse. For a long time the causes of the 1825 panic were difficult to discern; it was rooted, historians now believe, in a combination of overseas speculation, securities fraud, and rapid shifts in the money supply.[28] Regardless of its causes, its effects were quite clear to business owners like Martineau's father. Rapid inflation made the surplus goods at his factory all but worthless. The business declined alongside the health of its owner. Harriet's father died in 1826. An offer of marriage that might have saved her from poverty ended when her fiancé, John Worthington, became mentally unstable and died the next year. By 1829, the last remnants of the family business had dissolved. "I, for one, was left destitute," Martineau wrote later, "—that is to say, with precisely one shilling in my purse."[29] Because of a longstanding disability— Martineau, for reasons unknown, began going deaf in her teens—she could not fall back on teaching or working as a governess. Instead, she followed the working-class route of taking in needlework for money, supplementing that meager income with whatever she could earn by her pen.

To most people this turn of events would be devastating. Instead, Martineau would recall it as the period when her life "burst suddenly into summer."[30] Without a shred of middle-class respectability left, she could shamelessly take up the kind of writing she thought most necessary to the world. Her extensive reading and her own experience at the mercy of financial cycles convinced her that the average citizen would benefit from a better understanding of the principles of political economy. Those principles were typically expressed in dull academic language inaccessible to nonspecialists. Martineau was certain, however, that she could make such ideas comprehensible if she wove them into stories. Late at night, when the rest of her family slept, Martineau began writing fictions that showed how individual fates were determined by a combination of personal decisions and socioeconomic contexts.

The resulting tales, collected under the title *Illustrations of Political Economy,* were an instant success. They transformed the poor, unknown Martineau into a national celebrity and a household name. Yet they are rarely read today. Literary scholars have long dismissed them as financially successful but artistically insignificant. Critics cast the *Illustrations* as clunky and didactic at best, and often as something more nefarious: propaganda pieces serving the class of newly wealthy and powerful industrialists.[31] The tendency to downplay the significance of the *Illustrations* began with Martineau herself. In an autobiographical obituary that appeared days after her death, she insisted that "there is no merit of high order in the work. . . . It popularized, in a fresh form, some doctrines and many truths long before made public by others."[32] But the broadly derisive attitude toward the *Illustrations* is more specifically traceable to the profound animus many people feel toward one man in particular: Thomas Robert Malthus. While the *Illustrations* drew on the writings of Adam Smith, James Mill, David Ricardo, and other prominent political economists of the era, it was Martineau's status as an acolyte of Malthus that made her the target of attacks even in her own time. Detractors were not wrong in seeing Malthus as a principal influence on Martineau's work. As one of Martineau's friends observed, "There is a sense in which whoever teaches us any thing may be called our master. If any one in this sense was hers, I should have said it was Malthus."[33]

Ironically, it is precisely Malthus's influence that makes the *Illustrations* important from the perspective of intellectual history. Even when

Martineau's tales do not explicitly teach Malthusian lessons about population control, they fold Malthusian ideas into the underlying form of their storytelling. In story after story, Martineau constructs her narratives around the Malthusian idea that exchanges of matter and energy are the basis of interdependent communities—and that small changes in such dynamics can have dramatic, sometimes unpredictable effects as they ripple across interspecies networks. Whereas Malthus emphasized these ripple effects in a few key places in his *Essay*, Martineau used the *Illustrations* to expand them into elaborate ecological plots: sequences of cause-and-effect relationships that highlight the surprising ways in which human, animal, and vegetal fates are inextricably entwined.[34]

"Cinnamon and Pearls," the story that left Susan Darwin dreaming of Ceylon, is a case in point. It begins with Marana and Rayo, a young Sinhalese couple living on the island's northern coast. Both of them hope to marry as soon as Rayo can afford Marana's bride price. Their frustrated marriage plot is a product of the equally frustrated relationship between the Sinhalese and the natural resources of their island. Martineau is at pains to emphasize the natural wealth of Ceylon (now Sri Lanka), a "luxuriant region . . . where Nature seemed to intend that all things should flourish."[35] That wealth is not simply an abstract backdrop. It is materially present in the community of nonhuman creatures distributed around the island. The north is rich in chanks (locally prized, sacred shells that live in shallow beds off the coast) and pearls (global commodities drawn from banks farther out to sea). The jungles of the island's south abound with salt deposits, coconuts, peppercorns, cardamom, and cinnamon trees. But the proper relationship between the Sinhalese—"the natural owners of all the native wealth of their region"—and their environment is disturbed by a series of interlopers affiliated with British imperial forces.[36] The plot follows Marana and Rayo as they try to harvest assorted resources to accumulate wealth, only to be stymied by greedy foreign corporations and inefficient imperial bureaucrats. In short, the fate of Marana and Rayo hangs on a series of vexed engagements with the nonhuman world around them. These engagements, in turn, are mediated by various inhumane institutions that shape the couple's disastrous decision-making.

The setting of "Cinnamon and Pearls" lends it an exotic Orientalism that may help explain Susan Darwin's heightened awareness of its landscapes. In fact, as the environmental historian Richard Grove has shown, "the seeds of modern conservationism developed as an integral part of the

European encounter with the tropics," because Western excitement over non-European flora and fauna rendered the disappearance of such exotic resources more visible and more troubling than any analogous damage done at home.[37] As Jonsson has shown, however, Enlightenment concern over mismanaged natural resources was not confined to the tropics. Similar worries arose on Britain's own shores. In particular, the Scottish Highlands and islands provided the backdrop for many proto-ecological thought experiments; the area had a long history of unsuccessful attempts at land "improvement," failures that "gave rise to precocious anxieties about environmental degradation and resource scarcity."[38] Malthus wrote such concerns into the second edition of his *Essay*, which "singled out the Scottish Highlands and Hebrides as the region most burdened by overpopulation in Great Britain."[39] Martineau, in turn, borrowed these stark settings for the novellas she used to illustrate Malthusian themes most explicitly. So while Malthus's early ecological insights underlie the plots of many of the *Illustrations* tales, his demographic doctrine takes center stage in "Weal and Woe in Garveloch" (1832), a story set in a fictionalized version of the Hebrides.

Martineau characterizes Garveloch as an especially useful laboratory for demonstrations of Malthusian principles because of its isolation and limited resources. "[T]heir society was not like the population of an overgrown district," she remarks, "where there may be mistakes in ascribing effects to causes, and where the blame of hardship may be laid in the wrong place. The people of Garveloch might survey their little district at a glance, and calculate the supply of provision grown, and count the numbers to be fed by it, and by this means discern, in ordinary circumstances, how they might best manage to proportion their resources to labor and food."[40] This apparently straightforward Malthusian math is belied, however, by the ecological plotting that structures the story. As the narrator elaborates the material relationships binding islanders to herring, shellfish, grain, seaweed, potatoes, and rainstorms, Garveloch proves to be anything but a closed and calculable system.

The island's troubles begin after almost a decade of prosperity. Awakened to the abundance of unexploited resources around the Hebrides, a fishing company has bought rights from the local laird and built up an industry that transforms the landscape. Disaster strikes when a set of interrelated developments in the community—an abundance of children, an influx of unemployed immigrants, sabotage between jealous fishermen, and the replacement of traditional grains with less labor-intensive potatoes—suddenly cohere into a crisis. An unusually cold and stormy summer ravages the sensitive potatoes and drives away the shoals on which the fishery depends. Deprived of their staples, the locals scavenge shellfish and seaweed to survive, while opportunistic military recruiters swoop in to lure able-bodied men away and ship them off to war. Many islanders do not live to see another prosperous summer.

At a broad thematic level, the story tries to illustrate how Malthusian population theory should inform individual decision-making. "The increase of population is necessarily limited by the means of subsistence," Martineau flatly concludes in the "Summary of Principles" that follows the tale: "By bringing no more children into the world than there is a subsistence provided for, society may preserve itself from the miseries of want."[41] A smaller population might sometimes offer one way of reducing competition for limited resources, but Martineau's straightforward lesson about calculated sexual restraint does not map onto what actually happens in the novella. Instead of demonstrating the success that follows from chastity, foresight, and a careful assessment of resources, her ecological plot highlights just how contingent and incalculable the fate of even the most insular community can be.

Though the islanders inhabit what seems like a closed system, the story actually reveals them to be caught within intertwining natural and social networks that encircle the entire globe. The timing of a revenge plot in the story, for example, hinges on the staggered fishing seasons in the Hebrides, which in turn reflect the oceanic migrations of "herring shoals . . . and the cod which follow to make prey of them."[42] In this mesh of unpredictable entanglement, prudence may generally pay off and improvidence may be punished. Still, no characters come across as central or agentic in any straightforward way—they cannot, because their paths hinge on an astonishing array of intermediaries, both human and nonhuman.[43] Martineau even admits as much. Interspersed with her emphasis on the simple arithmetic of sustainability is another message that highlights the utter unpredictability of life. As she remarks at one point, individual successes or failures lie open to a sprawling set of happenstance events that may include "[a] single bad season, the opening of a few more fishing stations, a change in the diet of the West India Slaves,—any one of these."[44]

Even the stories not explicitly devoted to Malthusian lessons preserve the essentially Malthusian technique of tracing flows of food and resources to reveal the open, precarious web of interconnection on which human communities depend. These plots use literary form to teach a kind of ecological literacy, encouraging readers to look beyond conventional distinctions between human and nonhuman—the social and the natural—by rejecting analogous literary distinctions between foreground and background. The novellas repeatedly underscore the importance of learning to see through peaceful landscapes to the communal relations that structure them.

Some tales, such as "Sowers Not Reapers" (1833), explicitly emphasize the need for this kind of ecological literacy. Here, Martineau takes pains to highlight the danger of treating the setting as a stable, plot-free space surrounding society. The title of the story's first chapter, "Midsummer Moonlight," might seem to promise the kind of idyllic scene-painting that opens so many nineteenth-century fictions.[45] Martineau begins instead with a rejection of the peace and quiet her title promises, gesturing toward a world of hidden but violent material exchange permeating the landscape. "During the nights we speak of," she writes, "repose did not descend with the twilight upon the black moors of Yorkshire, and the moon looked down upon something more glittering than the reflection of her own face in the tarns of Ingleborough." The moonlight glinting across this landscape comes not from still waters but from the newly honed knives traded among the field laborers who are preparing a rebellion against their wealthy employers: "Some of the polished and sharpened ware of Sheffield was exposed to the night dews in the fields, and passed from the hands of those who tempered to the possession of those who were to wield it."[46] The coming class warfare remains invisible to "those sober people who shut themselves in" and write off the night landscape as a calm, uninteresting periphery to human activity.[47]

Several characters watch this roiling rebellion from spots of concealment around a dry spring, united by their shared desperation for a trickle of water that never comes. The second chapter follows these characters back to their village, examining the effects of that water when it finally arrives in the unwelcome form of thunderstorms. These storms are not some symbolic set piece intended to reflect the political storms brewing in England. Instead they provide an opportunity to carefully examine the diverse socioeconomic impacts of a sudden downpour on the lives and livelihoods of a range of interconnected entities, including "horned sheep"; "the shepherd"; "young plantations, where a thousand saplings stood, dry enough for firewood"; the local estate's "[t]rees of loftier growth"; "[e]very washerwoman . . . who happened to be pursuing her vocation that day"; "the millers and their men"; "careful housewives"; "thatch or tile" roofs;

"the ducks"; and "the straggling oats."⁴⁸ The climatological event gathers all these beings together, connecting each to each, but—just as in Malthus—not in a blithely holistic fashion. As it turns out, the entire plot of "Sowers Not Reapers" unfolds from the discrepant effects of the same wind, rain, and fiscal policies on this diverse catalogue of creaturely life. It's a complex plot with many actors and agents, only a few of them human.

But that is precisely the point. Martineau's writing uses narrative to follow the consequences of actions as they ramify outward across groups that include human and nonhuman beings. In the process, she shows how material exchange draws species together into interdependent communities whose dynamics were only beginning to be understood. The habit of consigning certain members of such communities to the category of setting or background obstructed this dawning understanding, making comprehensive, considered decision-making all but impossible. A more ample vision was necessary. Martineau drives this idea home near the end of "Sowers Not Reapers." There, the local squire's decision to cut down his stately oaks to pay for his children's schooling prompts considerable discussion of the transformations underway in what others might write off as mere scenery.

Martineau praises those who learn to read or see changes in the landscape and denigrates anyone who cannot. "The woodmen who sat on the fallen trunks thought little, while enjoying their meal and their joke, of all that was included in the fact of these trees having fallen," she observes.⁴⁹ Properly interpreted, however, all the features of a distant landscape reveal not peaceful repose but their own subtle parts in an extended narrative arc. The right kind of reading can connect the dots and identify changes cascading through this network of humans, animals, plants, and machinery: "[Y]et others had eyes wherewith to look beyond the green slope where they were sitting, and to mark signs of the times in whatever they saw;—the whirling mill, with one or two additional powdered persons on the steps, or appearing at the windows;—the multiplication of the smokes of Sheffield;—the laden lighters below Kirkland's granaries;—Anderson's fields, waving green before the breeze;—sheep and cows grazing where there was to have been corn;—and, above all, Chatham [a local stone-cutter] taking his way to the accustomed quarry, in a very unaccustomed manner."⁵⁰

The ability to read story where others see only serene, unchanging scenery is the basis of ecological literacy. As Timothy Morton explains, ecological awareness more or less obliterates our notion of background, scenery, or setting. "Since everything is interconnected," Morton observes, "there is no definite background and therefore no definite foreground."[51] Martineau's emphasis on this kind of literacy foreshadows the twentieth- and twenty-first-century work of May Theilgaard Watts, Tom Wessels, Robert Macfarlane, and other environmental activists who have emphasized how placid landscapes are actually sites of dynamic histories—histories that can be read and written by observers trained in the right kind of ecological interpretation.[52] None of these ecologically savvy writers would be likely to name Harriet Martineau as a major influence on their thinking. Nevertheless, their notions of ecological literacy owe a debt to her, because Martineau helped pass this mode of reading for interspecies interconnection from Malthus onto its most important advocate in the life sciences: Charles Darwin.

INTERCONNECTION IN DARWINIAN NATURAL HISTORY

Susan Darwin wasn't the only person to introduce Harriet Martineau's work to her influential brother. Sailing along the coast of South America aboard the *Beagle*, Charles was far removed from the craze for Martineau's tales that was sweeping the British reading public. Within two weeks of Susan's note, however, his sister Caroline independently shipped several numbers of the *Illustrations* to her brother. "Miss Martineau," Caroline explained, "is now a great Lion in London, much patronized by Ld. Brougham. . . . [E]verybody reads her little books."[53] We cannot know for certain whether Charles read them at the time, but there is every reason to believe he did. They were among the only known works of fiction included in the *Beagle*'s library—and while Charles did not bother to record most of his literary reading, he adored fiction throughout his life.[54] In his old age he confessed to having lost all taste for poetry, music, and painting, even as he remained an inveterate novel reader. Novels, he reflected in his *Autobiography* (1882), "have been for years a wonderful relief and pleasure to me, and I often bless all novelists. . . . I like all if moderately good, and if they do not end unhappily—against which a law ought to be passed."[55] It is hard to imagine that Charles abstained from this lifelong pleasure aboard the *Beagle*, where Martineau's novellas were literally delivered to his door.

Regardless of his reading decisions, Darwin had ample opportunities to converse with Martineau in the years to come. Her literary success soon shuttled her into the same parlors and drawing rooms as the respectable Darwin family. When Charles returned from circumnavigating the globe and Martineau arrived home from an extended tour of the United States, the two came to know each other well—so well, in fact, that Charles and his family privately expressed concern that the domineering Miss Martineau was angling for a marriage offer from Charles's bashful brother Erasmus. The family's fears turned out to be both ungenerous and unfounded. But the years to come threw Martineau and Darwin together with some regularity, including at a dinner party where Charles "had a very interesting conversation with Miss Martineau" in May 1838. Only a few months later Darwin famously "happened to read for amusement Malthus on *Population*," and as a result he developed his revolutionary theory of natural selection that explained, for the first time, how all living beings could evolve from simple forms.[56] Darwin never clarified how he came to read Malthus for amusement, but scholars speculate it was likely the effect of Martineau, who evangelized on behalf of Malthus to the end of her life.[57]

We may never know with total certainty which of these encounters led Darwin to develop his ardent devotion to Malthus, a devotion that rivaled or even exceeded Martineau's.[58] What is clear, however, is that these two influential thinkers of the Victorian era developed a passion for the same unloved economist and that this passion had a profound effect on how they saw and described living communities. Most accounts of the importance of Malthus to Darwinian thought focus only on the theory of natural selection. In the standard telling, Malthus's bleak vision of an overpopulated world helped Darwin see that competition for resources could cause particularly well-adapted creatures to survive while others died off. This differential survival rate would eventually spread useful adaptations through the population, enabling evolution to take place over the course of many generations. But Malthus also left his mark—perhaps through Martineau's novellas, or perhaps more directly—on the way Darwin saw community dynamics playing out over the shorter timescales that define ecology.

Indeed, while *On the Origin of Species* is best known as the text in which Darwin explains the theory of natural selection so central to modern biology, it played an equally pivotal role in promoting the view of interspecies relations central to ecology. The book features a number of stunning anecdotes about biotic communities that closely resemble, in compressed form, the ecological plots featured throughout Martineau's *Illustrations*. The best example of this resemblance appears in a famous passage from chapter 3 ("Struggle for Existence") of the *Origin*. There Darwin establishes the basis for both natural selection and ecological ideas by narrating interactions "showing how plants and animals, most remote in the scale of nature, are

bound together by a web of complex relations."⁵⁹ In one such micronarrative, he begins with a set of observations of grazing cows devouring evergreen saplings on English heathland, proving that "cattle absolutely determine the existence of the Scotch fir" in such scrubby landscapes (60).⁶⁰ Building on this relationship between grazers and tree growth, he observes that "in several parts of the world insects determine the existence of cattle," because insects—flies, mostly—kill many calves shortly after birth by laying eggs in their tissues (60–61).⁶¹ But such flies, in turn, are eaten by birds. This strange chain of relationships means that a change in the population of "certain insectivorous birds (whose numbers are probably regulated by hawks or beasts of prey)" would lead to a change in the population of insects, leading to a change in the fates of cattle, and, finally, to a dramatic transformation in the vegetation of the landscape itself—"and so onwards in ever-increasing circles of complexity" (60–61).

Another half-speculative, half-empirical chain of interspecies relations follows shortly thereafter, illustrating just how widespread and mindboggling such relations may be. Darwin's observations led him "to believe that humble-bees are indispensable to the fertilisation of the heartsease (Viola tricolor). . . . [and that] humble-bees alone visit the red clover (Trifolium pratense)" (61–62). The preponderance of violets and clover in English meadows thus depended on the population of bumblebees.⁶² The bumblebee population, however, "depends in a great degree on the number of fieldmice, which destroy their combs and nests" (62). The field mouse population was heavily affected by the number of cats around to prey on them. Taken together, that chain of interdependence leads Darwin to a vision of communal relations so unexpected it shocked even the great naturalist himself, who expresses his surprise with an uncharacteristic exclamation point: "Hence it is quite credible that the presence of a feline animal in large numbers in a district might determine, through the intervention first of mice and then of bees, the frequency of certain flowers in that district!" (62).

The startling depictions of interspecies communities in these micronarratives scattered through Darwin's book also struck the German naturalist Ernst Haeckel as revolutionary—so revolutionary that he saw them forming the foundation for an entirely new field of scientific study. In recognition of this watershed moment he coined a new word, *Oekologie,* to describe the investigation of "those complicated mutual relations which Darwin designates as conditions of the struggle of existence."⁶³ The dependence of this new discipline on a particular mode of storytelling escaped Haeckel. It becomes clear, however, when the writings of Darwin and Martineau are read in tandem. Both utilize a strategy of storytelling that traces material processes of cause and effect across conventional boundaries of space, class, and species to reveal the webs of reciprocal relations that structure living communities. And in both cases that strategy is linked to their formative encounters with the work of Malthus. The precise details of the intellectual traffic between these figures—what Malthusian insights Darwin learned from Martineau, and what he got from Malthus directly—are finally less important than the storytelling strategies they shared. What Darwin and Martineau both ultimately derived from the *Essay* was a narrative form that afforded ecological thought, a strategy of storytelling that made it possible, even easy, to see how human beings were materially inextricable from the nonhuman world around them.[64]

As was the case with Malthus, the stories of interconnection that Darwin and Martineau produced did not radically alter their senses of ethics. Each thinker operated within certain professional standards, and those standards shielded them from thinking through the moral or sociopolitical consequences of their ecological storytelling. Thus while Darwin, a natural historian, indulged in aesthetic admiration for the complexity of interspecies relations, the ecological anecdotes in the *Origin* remain descriptive rather than prescriptive. As a scientific observer, he abstains from any serious engagement with the moral repercussions of interconnection. In his thought experiment about how a decreased number of cats in England would lead to more mice, fewer bumblebees, and fewer violets and clover, for example, Darwin himself never posits whether such a decrease would be a good thing or a bad one.

Martineau, by contrast, had no problem freighting her tales with moral teachings. Indeed, the morals of her stories are conveniently catalogued in the final pages of each installment. But those teachings are reductive in comparison with the stories themselves. The lessons merely reiterate the ethical prescriptions of classical political economy, with its admiration for free markets structured around prudent self-interest. Martineau's fictions thus fail to grapple with the radical interspecies entanglement that her stories illustrate so skillfully. She downplays the complications of the world she depicts, highlighting instead a set of comparatively simple economic principles drawn from Smith, Malthus, and Ricardo.[65] Critics have long decried this feature of the *Illustrations*, arguing that Martineau's adherence to the dictates of political economy compromises her work by turning "complex problems into happy fables ... [written] in a mechanistic prose analogous to the immutable laws governing the ultimately benevolent workings of such economic systems."[66]

In the end, then, all the early practitioners of ecological plotting avoided the ethical quandaries arising from interconnection. Malthus, Martineau, and Darwin each told stories to illustrate how material exchange knit communities together in surprising ways—ways that had the power to redefine who counted as a member of the community and how self-interest should be understood. Yet they mostly shied away from such redefinition, sticking instead to the ethical stances endorsed by other practitioners of political economy and natural history. The task of thinking through such redefinitions and revaluations was left to other authors who did not share such disciplinary loyalties. Those authors were not naturalists or political economists at all. They were novelists.

The fiction writers who followed in Martineau's footsteps admired her use of storytelling to explore entangled fates, and they borrowed from her freely. But they did not share her confidence that a providential moral order would emerge from some combination of free markets, foresight, and the innate instinct of self-preservation. As a result, they had to face the disorienting ethical uncertainties arising from an awareness of ecological interdependence—and they had, somehow, to overcome them.

CHAPTER 2

The Price of Interconnection

The year was 1848. All across Europe, society teetered on the brink of collapse. In February the last French king, Louis Philippe I, fled from the increasingly violent protests surging through Paris. His retreat to Switzerland set the stage for the presidential elections that would usher in the Second Republic. Just days earlier, a German exile named Karl Marx had published a pamphlet, *The Communist Manifesto,* that declared class struggle to be the unifying theme of the history of civilization. It was certainly the unifying theme of that fateful year. Inspired by their neighbors in France, the continent's loose consortium of German-speaking duchies and principalities soon boiled with unrest as citizens rejected their appointed aristocratic rulers, demanding constitutional reform and a united German republic. Local leaders in Sicily, Sardinia, and the northern Italian peninsula rose up and fought the Bourbon and Habsburg Empires that controlled them. In Hungary, a newly representative government revolted against the Austrian monarchy, triggering a revolutionary war.

The British ruling classes watched these events with a growing sense of disquiet. Compared to the radicals on the continent, the citizens of the first industrialized nation in the world seemed relatively peaceful. Newly prosperous businessmen and factory owners had successfully pushed for some voting reforms a decade earlier, a move that eased the social tensions inspiring so many revolutions abroad. The Corn Laws that kept wheat prices artificially high for decades had been repealed in 1846, freeing up trade and reducing the starvation that had prevailed during a decade popularly dubbed the "hungry forties."

Yet even in Britain, anger and internal division persisted. A mass movement for workers' rights known as Chartism refused to die down. In 1839,

1842, and 1848, millions of laborers came together to sign petitions calling for a more democratic voting process, working-class representation, and political accountability. Parliament refused to accept these petitions or even meet with the men who delivered them. The recurring agitation for the People's Charter touched off widespread debate around the relations between the rich and the poor in industrial nations, a topic the writer Thomas Carlyle called "the Condition of England Question."[1]

Political economy had already answered this question—in theory, anyway. Its acolytes touted their booming field as the study of prosperity itself. It was the science that finally explained the material ties that bound society together. In doing so it could help steer the community toward "the greatest happiness of the greatest number," in the words of Jeremy Bentham, one of the discipline's early advocates and enduring influences.[2] Popularizers like Harriet Martineau made this obscure field accessible to legislators and the general public alike. By midcentury the logic and dictates of political economy were broadly familiar and widely applied. Yet the new science had failed to deliver on many of its promises. The greatest number of people still did not appear to be enjoying very much happiness. The tension between workers, middle-class manufacturers, and aristocratic landowners refused to go away. Something was missing. And so a new crop of social chroniclers arose after the pattern of Martineau, intent on filling in the gaps of previous explanations to provide the reading public with a clearer and more comprehensive understanding of society. But these writers were not political economists, or academics, or even essayists. They were authors of popular fiction.

During the 1840s and 1850s, a groundswell of novels appeared that strove to paint more realistic pictures of industrial society than the economic accounts that now wielded such power. Known collectively as the industrial novels, the Condition of England novels, or the social-problem novels, they portrayed lived experiences of poverty and strife that undermined the self-assured pledges of political economists. Some of the authors of these works—including Charles Dickens, Benjamin Disraeli, and Charlotte Brontë—were already established in literary circles when they turned their attention to the ravages of industrial life. Other writers, such as Charles Kingsley and Elizabeth Gaskell, were comparatively new to the literary scene. Taken together, their works helped solidify a tradition pioneered by Harriet Martineau: the tradition of using narrative fiction to offer sweeping pictures of the material networks that drew society together. The industrial novels thus set the stage for the ambitious portrayals of social life that became characteristic of the golden age of the English novel.[3]

If the early social-problem novels are notable for the ways they borrowed from the ecological plots that preceded them, they are equally notable for the ways they broke precedent. These industrial fictions were written at the start of a major rupture in modern ways of imagining interdependence. The storytelling mode they used was on the verge of splintering into three distinct and competing approaches to understanding communal life. By the late nineteenth century, the dazzling intellectual constellation once formed by Malthus, Darwin, and Martineau would be all but invisible, their guiding lights partitioned off and assigned to the distinct fields of economics, natural science, and realist literature. The social-problem novels preserve the first signs of such division and discord in their pages. Again and again, these early Victorian works express frustration at the failures of both political economy and natural history as techniques for understanding the living world.

What concerned the social-problem novelists was, in essence, a crisis of valuation. Political economy and natural history offered powerful ways to describe the many beings in the world and the relationships that bound them together. Even before the publication of Darwin's *On the Origin of Species*, these two fields seemed to be cohering around a unified model of the world as a single, seamless marketplace. The earth was a network of exchange where everything was interconnected and anything might be traded for or transformed into anything else—for the right price. This emerging worldview had a kind of simple, egalitarian appeal. It offered a newly universal system that overthrew archaic hierarchies based on the mystique of categories like nobility and heritage. It provided techniques for making all things exchangeable, assigning them calculable values according to the price they would fetch in the marketplace.

But for the industrial novelists there was something sinister about this new equality. Its vision of endless interconnection and frictionless free-market exchange upended traditional distinctions not just among people but also between people, animals, plants, and objects. Traditional relationships included traditional protections—codes of behavior that involved valuing farmhands more than the suckling pigs they raised, for example—and such protections no longer held sway. They were eclipsed by a more logical but

more merciless system that assigned rank and value almost exclusively in terms of monetary profit. For the more extreme partisans of political economy, all the world was a marketplace, and all the men and women in it merely commodities. They had their profits and their costs, just like any other good or resource. Other ways of understanding community were consigned to premodern ages of unreason.

This laissez-faire approach to interconnection did not sit well with a new generation of Victorians. Coming of age at a time defined by the proliferation of increasingly cheap and plentiful printed matter, the social-problem novelists took to their pens to sound a general alarm about the dangers of unchecked economic logic. The early examples of this subgenre provide a useful case study in the larger upheavals of the era, the upheavals that left fissures between the emerging fields of economics, ecology, and the realist novel. And no novel serves as a better example of this decisive turn in Victorian letters—and its significance for larger understandings of interconnection—than Elizabeth Gaskell's *Mary Barton* (1848).

Debuting to great fanfare in a year rocked with revolutions, *Mary Barton* demonstrates just how much the midcentury industrial novelists owed to the ecological plots popularized by Martineau a decade earlier. Yet *Mary Barton* rejects the individualism of political economy, deriving a very different ethical conclusion from the same premises. Notably, Gaskell includes both industrialists and naturalists among her characters, using them to demonstrate the ethical inadequacy of both fields when confronted with the dilemmas of interconnection. Gaskell's seemingly demure novel thus does a striking job of articulating a widespread feeling among authors at the dawn of the Victorian era. Economic thinkers may have revealed the complex dynamics tying humans and nature together, but they had proven incapable of responding to such interdependence with the moral attention it required. The other scholars who might help—naturalists who studied the nonhuman portions of such networks—were so caught up in acquiring, trading, and cataloguing specimens that they put little effort toward ethical reflection. As Gaskell and her peers saw it, the public needed help moving from an abstract appreciation of resource networks to the ethical upshot of interconnection. People needed to develop the habit of understanding themselves as nested within such networks, as morally responsible for the successes and failures of their fellow creatures. It was a matter of reimagining society as a whole. And rather than idly hoping such a reimagining would happen on its own, the social-problem novelists set to work putting this new social vision to paper and placing it under the noses of the reading public.

ELIZABETH GASKELL AND THE ETHICS OF INTERDEPENDENCE

Who was responsible for *Mary Barton*? When the book first appeared to widespread acclaim in 1848, no one was certain who wrote it—but Maria Edgeworth had a hunch.[4] Edgeworth had once numbered among the most successful novelists of the Romantic era, her works read and discussed in the same breath as those by Jane Austen and Sir Walter Scott. By the 1840s her professional peak was well behind her. Still, it was with a certain insider knowledge that Edgeworth guessed *Mary Barton* was the production of an experienced author: in her opinion, the novel came from the already prolific pen of Harriet Martineau.[5]

As it turned out, Edgeworth's guess was incorrect. The author of *Mary Barton* was actually Elizabeth Gaskell, a minister's wife and first-time novelist living in Manchester. Nevertheless, Edgeworth was right to detect traces of Martineau in Gaskell's debut. The economic popularizer was a formative influence on *Mary Barton* and those other early Victorian fictions that came to be known, collectively, as the industrial novels. The authors of these works frequently singled out Martineau for praise even when they disagreed with her. Charles Kingsley counted her as one of the "good and wise people" of the era. George Eliot lauded her as a *"trump*—the only English woman that possesses thoroughly the art of writing."[6] In Martineau's earliest tales and in the *Illustrations* that followed, the abolitionist Maria Weston Chapman identified "the first examples of a new application of the modern novel. To the biographical and the philosophical novel, the descriptive and the historical novel, the romantic and the domestic novel, the fashionable and the religious novel, and the novel of society, was now to be added the humanitarian or novel of social reform. These tales are the pioneers . . . of the multitudes of social-reform novels that have since followed, up to the time of Mrs. Gaskell and Mrs. Stowe."[7]

Literary historians today affirm Chapman's nineteenth-century assessment. As Eleanor Courtemanche observes, Martineau's economic tales possess a strong claim to being "the first secular industrial novels." Their popularity, combined with their attempts to map the contours of entire communities, laid the groundwork for later social novels by establishing a wide readership that learned to accept "fiction's power to reveal social truth."[8] What tied later realist writers to Martineau was not simply a commitment to social reform, however. They also relied on her innovations in storytelling.

Realist writers after Martineau often borrowed the ecological plot structures she popularized, using them as the scaffolding for their own tales of communal interdependence. *Mary Barton* offers an unusually direct and clear-cut example of such borrowing. Gaskell and Martineau had been friends for over a decade before *Mary Barton* was published, and scholars have long acknowledged that *Mary Barton* draws on one of the tales in Martineau's *Illustrations*, "A Manchester Strike" (1832), for source material.[9] These intertextual echoes and personal intimacies attest to the influence Martineau had on Gaskell—but as Maria Edgeworth noted, that influence is perfectly legible within Gaskell's novel itself.

Gaskell's storyline is a direct descendant of Martineau's proto-ecological *Illustrations*. Her plotting shows how material exchanges bind human communities and the natural world together into unexpected networks of interconnection, networks that erase easy distinctions between foreground and background, society and environment, human and nonhuman. The ecological form of *Mary Barton* can be hard to appreciate, however, due to the subtlety of the novel's plotting and its overwhelmingly urban setting.[10] The story focuses on the struggles of a small circle of working-class friends and family in the city of Manchester. It centers especially on the young dressmaker Mary Barton and her father, John, an unemployed factory worker. As John is drawn deeper and deeper into the political turmoil surrounding Chartism and union organizing, Mary finds herself caught in a love triangle between her working-class friend Jem Wilson and Harry Carson, a rich industrialist's son. When Carson is shot with a bullet from Wilson's gun, it falls to Mary to prove Jem's innocence before he is executed for the crime.

Aside from a brief jaunt in the countryside that opens the novel and a glimpse of the Canadian suburbs that concludes it, Gaskell's entire story takes place within cities—first the smoky, sewage-riddled streets of Manchester and later the hurried maritime hub of Liverpool. Nevertheless, connections between human society and the nonhuman world structure every element of the novel's plotting. Their importance is most obvious near the dramatic climax, when Mary discovers that a witness who could provide an alibi for Jem is aboard the *John Cropper*, a ship that left Liverpool the night before. Just when she believes all is lost, Mary learns that "there's sand-banks at the mouth of the river, and ships can't get over them but at high water; especially ships of heavy burden, like the *John Cropper*."¹¹ Mary needs to get aboard a boat herself to see if she can catch up to the ship before it departs for good.

At a purely mechanical level, the ship-chase scene functions as a sensational plot device, a clever mechanism for increasing the suspense of the narrative. From a sociopolitical perspective, however, the device has a more profound significance. It sets up a situation where Jem's life hangs in the balance, and what will determine his fate in the end is not his character, or his actions, or Mary's actions, or any other agency that is exclusively human. The story hinges on the question of whether Mary's actions are properly timed in relation to a shifting conjunction of tides, wind, lading, and the unpredictable wanderings of the sediment deposits that form at the mouth of the Mersey estuary. Melodramatic as it may seem, then, this climactic moment is also a climatic one. It is grounded in geophysics. It drives home just how much Gaskell had internalized the networks of natural and socioeconomic factors that typified Martineau's ecological plotting.

Less dramatic but no less significant are the concatenations of social, economic, and natural occurrences that shape Mary's earlier experiences in the Manchester slums. Sometimes these interlocking elements are described only in the most abstract terms. So, for example, John Barton's anger at economic inequality simmers beneath the surface as long as he and his family possess a certain level of financial stability. After describing "the hoards of vengeance in [John's] heart against the employers," Gaskell explains how a strong economy kept his vengeance in check: "But now times were good; and all these feelings were theoretical, not practical" (25). It will take some alteration in the broader system before John is spurred to action. When it comes time to describe such alterations, however, Gaskell works hard to pinpoint how natural phenomena, economic structures, and human doings dovetail to make history happen. One of the most important instances of this finely tuned ecological plotting takes place early in the text, when a paragraph of apparently innocuous scene-setting turns out to be crucial to the entire novel that follows:

> It was towards the end of February, in that year, and a bitter black frost had lasted for many weeks. The keen east wind had long since swept the streets clean, though in a gusty day the dust would rise like pounded ice, and make people's faces quite smart with the cold force with which it blew

against them. Houses, sky, people, and everything looked as if a gigantic brush had washed them all over with a dark shade of Indian ink. There was some reason for this grimy appearance on human beings, whatever there might be for the dun looks of the landscape; for soft water had become an article not even to be purchased; and the poor washerwomen might be seen vainly trying to procure a little by breaking the thick gray ice that coated the ditches and ponds in the neighborhood. (43)

As in Martineau's *Illustrations,* this atmospheric background will soon come to the fore, erasing easy distinctions between plot, character, and setting. In the midst of this bleak winter Mary's friend Margaret Jennings has accepted a last-minute, all-night job of sewing funeral attire for the Ogdens, a local family. Mary decides to sit up and help her friend with the stitching, which leads them into a thoughtful, wide-ranging conversation on everything from the expense of mourning rituals to the eyestrain of doing needlework in dim rooms. Their discussion is interrupted suddenly by news of a fire at the Carsons' mill. As John Barton explains, the fire is certain to be an unusually explosive event: "[A] rare blaze, for there's not a drop of water to be got. And much [the] Carsons will care, for they're well insured, and the machines are a' th' oud-fashioned kind. See if they don't think it a fine thing for themselves" (49). Mary and Margaret rush out to catch the spectacle, only to discover that Mary's childhood sweetheart Jem Wilson is in danger. They watch as he repeatedly braves the flames to rescue his father, who works in the mill.

Everyone is safe by the end of the night. Yet there is nothing neatly self-contained about the incident. It casts a long shadow, its consequences extending gradually across the community to increasingly devastating effect. In the short term, watching the blaze leaves both Mary and Margaret exhausted. Upon returning to their lodgings, Gaskell explains, "The work which they had left was resumed, but with full hearts fingers never go very quickly; and I am sorry to say, that owing to the fire, the two younger Miss Ogdens were in such grief for the loss of their excellent father, that they were unable to appear before the little circle of friends gathered together to comfort the widow, and see the funeral off" (55). Here Gaskell is making a lighthearted joke at the Ogdens' expense. The actual reason for the daughters' absence is that they lack proper mourning attire and so cannot appear at a memorial service without embarrassing themselves socially. The more official and decorous explanation offered to funeral guests is that excessive grief forced the bereaved daughters to shield themselves from the public eye. Gaskell humorously conflates the two, drawing attention to the mismatch between how we discuss the world socially and the more complex factors that contribute to how the world actually works.

Amusing irony aside, the absence of the Misses Ogden is worth lingering over precisely because it showcases the novel's intricate ecological methods. The young Misses Ogden cannot appear because they lack mourning attire. Traditionally, they would have fashioned something seemly themselves, but Mrs. Ogden has sent their mourning attire out because "th' undertakers urge[d] her on"—a symptom of an increasingly corporate funeral industry that goaded the grieving into spending more money to honor the dead (45).[12] The mourning attire ends up in the hands of Mary and Margaret, but they cannot finish it because they spend such a long time watching the fire. The fire blazes so long because the east wind rapidly spreads it along the mill, "which ran lengthways from east to west," and the water to put it out is frozen (49). And finally, the effects of these cultural, geographic, and meteorological factors are amplified by the idleness of the mill owners, who don't bother rushing for the fire engines because they know their insurance will cover the costs of rebuilding an updated, more efficient mill.

If the only product of this interlocking mechanism of natural, cultural, and economic factors were the absence of two minor characters at their father's funeral, it might not merit discussion. As it happens, though, that frozen February has a number of long-term consequences as well—effects that are easy to forget in such an intricate, compounding plot. The night of sewing worsens Margaret's encroaching blindness, which undermines her ability to support herself and her grandfather. At the same time, the fire that interrupts her sewing destroys the Carsons' factory, which changes the economic situation of the neighborhood. Because they enjoy a substantial financial cushion, the Carsons decide to take advantage of the fact that "trade was very slack" at the moment to reinvest in the infrastructure of their textile mill—a process that enables them to step away from business temporarily and devote themselves to "happy family evenings" and "domestic enjoyments" instead (56). Their workers don't share in the Carsons' prosperity, so they are forced to spend the furlough fighting off sickness and starvation. One steady, uncomplaining hand from the mill named Ben Davenport weakens from hunger and catches typhoid fever. When John Barton steps in to help Davenport in his final hours, the experience changes Barton forever. "When I see such men as Davenport there dying away, for

very clemming [i.e., starving], I cannot stand it," Barton rages (64). He decides the rich would respond differently if such deaths came home to them in the shape of their own children—"Han they ever seen a child o' their'n die . . . ?"—a theory he later tests by scheming with others to kill a mill owner's son (64).

Far-fetched as it initially sounds, then, every significant plot development in *Mary Barton* finally arises from the combined effects of temperature and wind direction as they come to bear on a particular set of social and economic relationships. This ecological plotting is subtle but pervasive. Where Gaskell differed from her friend Martineau was in her assessment of the moral implications of this interconnection. It was this kind of ethical disagreement that would finally drive a wedge between realist writers and the early economic and ecological thinkers who shared their general understanding of communal interdependence.

Following the prescriptions of political economy, each of Martineau's *Illustrations* had ended with a tidy list of lessons that readers should take from the tale in question. The discrepancy between these prescriptions and the moral feeling of a realist writer like Gaskell becomes clear in the case of "A Manchester Strike," the tale believed to have inspired parts of *Mary Barton*. At the conclusion of that story, Martineau enumerates how the miseries of factory workers could be lessened. The only remedy, she explains, is for workers to help themselves. "The condition of labourers may be best improved," she writes,

1st. By inventions and discoveries which create capital.
2nd. By husbanding instead of wasting capital:—for instance by making savings instead of supporting strikes.
3rd. BY ADJUSTING THE PROPORTION OF POPULATION TO CAPITAL.[13]

Martineau insisted, in other words, that only workers could reduce their own suffering and that the only ways they could do so were to innovate, to save money, to avoid unions, and (in all-capital letters!) to stop having so many babies. Her belief that these individual initiatives would be more effective than strikes or other collective action mirrored contemporary economic thought—in particular, the so-called Iron Law of Wages. Derived from a combination of Adam Smith's classical political economy, David Ricardo's theory of rent, and Malthusian population theory, the Iron Law of Wages asserted that payment for unskilled labor always naturally fell to the lowest level that would keep essential workers alive. Any apparent increase in wages must be superficial and temporary, because higher pay for workers would ultimately result in price or population increases that would erase all apparent gains.[14]

Under this model, the only hope for societal progress sprang from the individual. By inventing new technology, individual workers could acquire more capital for themselves. By saving money, they might lift themselves above the level of subsistence. And by implementing some form of birth control, they could lessen their expenditures at the same time that they diminished future competition from a growing number of young laborers entering the workforce. In short, "A Manchester Strike" and Martineau's other *Illustrations* stories saw nature and society coming together to produce a set of quasi-natural laws—the laws of economics—that constrained political possibilities and helped determine individual fates. If it was true that society obeyed such laws, the point was to discover them and work within them, not to denounce them. Fighting economic laws would be no more effective than fighting gravity. Martineau responded by suggesting individual strategies for navigating these laws to the greatest personal advantage.

Mary Barton and the other industrial novels recognized the power of political economy's interconnected accounts of society, but they refused the economists' jump from embodied, material relationships to abstract unbending laws. For these writers, the interplay of natural and social agencies did not fit together into a seamless piece of machinery, a mindless mill that operated in accordance with preordained rules to separate the savvy from the undeserving. Instead, the interplay of such agencies revealed a sprawling, complex community that required individuals to acknowledge previously unsuspected forms of obligation and moral duty.

Even the malfunctioning elements of this interdependent web deserved material and moral attention. Thus when John Barton sinks to his lowest point—unemployed, addicted to opium, beating the daughter who supports him, and agitating for violence against the middle classes—Gaskell refuses to let middle-class readers judge him or derive didactic lessons from his fate. She calls for sympathy instead, and she assigns anyone who would judge John significant responsibility for his fallen state. "The people rise up to life," she writes, in an analogy that equates the working classes to Frankenstein's monster. "They irritate us [middle-class citizens], they terrify us, and we become their enemies. Then, in the sorrowful moment of our triumphant power, their eyes gaze on us with mute reproach. Why have we made them what they are; a powerful monster, yet without the inner means for peace and happiness?" (165).

What makes John Barton sympathetic is, finally, his commitment to the well-being of the collective—his refusal to fall into the facile trap of self-interest to which political economy catered:

> John Barton became a Chartist, a Communist, all that is commonly called wild and visionary. Ay! but being visionary is something. It shows a soul, a being not altogether sensual; a creature who looks forward for others, if not for himself.
>
> . . . And what perhaps more than all made him relied upon and valued, was the consciousness which every one who came in contact with him felt, that he was actuated by no selfish motives; that his class, his order, was what he stood by, not the rights of his own paltry self. For even in great and noble men, as soon as self comes into prominent existence, it becomes a mean and paltry thing. (165–66)

John Barton may be wrongheaded, but at least he has the ability to see himself as part of something larger, to identify with and to use his powers for the greater good. The final moral vision of the novel is little more than an expanded version of John's class solidarity, a kind of social solidarity that extends across the entire community. "Distrust each other as they may," Gaskell insists in the final chapters, "the employers and the employed must rise or fall together. There may be some difference as to chronology, none as to fact" (166).

In *Mary Barton*, then, interdependence is more than a set of complicating laws that obstruct individual paths to success. Interdependence transforms the very ideas of individuality and success almost beyond recognition. It provides a reason to identify with others, to share wisdom and strategies for communal survival and prosperity. Even the smallest act of solidarity can have outsized effects. For example, when a young boy becomes invested in Mary's need to retrieve Jem's star witness before the *John Cropper* heads into the Irish Sea, the boy's simple act of identification converts Mary from despair to practical action:

> "Don't give up yet," cried the energetic boy, interested at once in the case; "let's have a try for him. We are but where we were, if we fail."
>
> Mary roused herself. The sympathetic "we" gave her heart and hope. (278)

The renewed formation of this "sympathetic 'we'" is the fundamental project of *Mary Barton* and its cohort of social-problem novels. The improved working conditions that Gaskell imagines at the conclusion of the book result not from economic trust in individual inventiveness, financial responsibility, and family planning but from a compassionate desire to prevent the spread of suffering through the community. In the end, even the hard-hearted Mr. Carson becomes a sort of reformer, his actions newly grounded in "the wish . . . that none might suffer from the cause from which he had suffered; that a perfect understanding, and complete confidence and love, might exist between masters and men; that the truth might be that the interests of one were the interests of all, and, as such, required the consideration and deliberation of all" (374).

In both its debt to economic narratives and its departure from them, *Mary Barton* is typical of the industrial novels as a group. At a very basic structural level, these works adopted an understanding of society as an extended network of material relationships that could be imagined and understood through the art of storytelling. Even the most right-wing social-problem novelists understood their project as healing division by foregrounding the connections between rich and poor. The rich and poor had begun to constitute "[t]wo nations," as the conservative Benjamin Disraeli influentially wrote in *Sybil* (1845), "between whom there is no intercourse and no sympathy; who are as ignorant of each other's habits, thoughts, and feelings, as if they were dwellers in different zones, or inhabitants of different planets."[15] Those who ignored the material reality of interconnection did so at their own peril. Take, for example, the case of the preening cousin in Charles Kingsley's *Alton Locke* (1850). This proud, self-interested heir finally dies of a strain of typhus fever first contracted by a sweatshop worker, then later by "the servant who brushed his clothes, and the shopman who had a few days before brought him a new coat home," because this disease bred of inequality could not finally be confined to the working classes; it passes easily along the supply chain from garment workers through the retail and service industries to the wealthy purchasers of industrial goods.[16] Later in the nineteenth century this sort of networked plotting would become so entrenched in realist social fiction that writers explicitly meditated on it in the pages of their own novels. Thus Dickens famously pauses his sprawling portrait of English society in *Bleak House* (1852–53) to whet his readers' curiosity about the surprising lines of material interdependence that link together to form the story he is telling: "What connexion can there be between the place in Lincolnshire, the house in town, the Mercury in powder, and the whereabouts of Jo the outlaw with the broom, who had that distant ray of light upon him when he swept the church-yard step? What connexion can there have been between many people in the innumerable histories of the world who from opposite sides of great gulfs have, nevertheless, been very curiously brought together!"[17]

This vision of networked collectivity originated in political economy. Nevertheless, the social novelists who employed it thought of themselves as dead set against that field and its lessons. At best they simply refused to engage with the subject. *Mary Barton,* for example, is relatively gentle in its departure from political economy. Gaskell attacks calculating self-interest and praises interpersonal ethics, but she never explicitly calls out political economists themselves for their wrongheaded theories of society. In fact she feigns total ignorance of the field despite a number of biographical facts that indicate an intimate acquaintance with it, including her educated upbringing, her friendship with Martineau, and the fact that her own father had published a series of essays on the subject.[18] "I know nothing of Political Economy, or the theories of trade," Gaskell announces rather unconvincingly in her novel's preface. "I have tried to write truthfully; and if my accounts agree or clash with any system, the agreement or disagreement is unintentional" (4).

Other industrial novelists would lampoon political economy more directly, often sniping at Malthus in particular.[19] Precisely because these attacks are so overt, studies of Victorian fiction have tended to emphasize novelists' dissent from prevailing economic opinion, their struggle "to challenge the modes of representation that political and social economists deemed adequate to contemporary woes."[20] But in noting such objections, it is vital to acknowledge the continued historical relations between the two kinds of writing. These relations are registered in the fields' shared imagination of society as a network of material connections.

When novelists argued that interdependence implied a need for fellow feeling and a renewal of moral duty, however, they ran into a difficulty that never troubled their economic predecessors. Political economy had revealed a web of connections that suggested humans relied on one another in unappreciated ways—but the webs of interconnection economists discovered did not stop at the species boundary. Malthus's insights were not merely economic; they were also inherently ecological. He emphasized how seemingly abstract calculations about the proportions of land given over to animal and plant life were, in reality, a matter of life and death for large swaths of humanity. Malthusian interdependence extended outside the bounds of the human community to include a wide range of other species entangled with humans in complicated ways. And when novelists insisted that such interdependence required a reconsideration of ethics, they created an unexpected problem for themselves. Any moral argument based on interdependence had to contend not only with the way human beings' fates are bound up with one another but with the way our fates are bound up with those of other creatures, too.

NATURAL RESOURCES AND HUMAN RESOURCES

One thing conspicuously absent from *Mary Barton* is an actual merry barton. The phrase "merry barton" literally means "happy barnyard," the sort of agricultural setting where established hierarchies among humans, plants, and animals are clear and untroubled—for humans, at least. Because the word *barton* has fallen out of use, the pun is easy to overlook. But it serves as a vital reminder of something deeper that is lacking from the worlds depicted in industrial fiction. Taken together, the social-problem novels struggle to find a reliable method of arranging their interconnected webs of humans, plants, and animals into neatly ordered systems built on moral certainties. In these early Victorian works the fantasy of a traditional, fulfilling rural life gives way to a chaotic modern jumble of inhumane industrialists, dehumanized workers, and a cast of animal characters with unnervingly human features.

These animalized humans and humanized animals are symptoms of a larger crisis of value that accompanied economic treatments of interconnection. Cultural historians have long acknowledged that social-problem novelists resisted the commodifying effects of economic thought, decrying its tendency to replace respect and fellow feeling between employer and employee with the quantitative valuations of the marketplace. But this ethical concern about the dehumanizing effects of political economy also coincided with a diametrically opposed concern: an anxiety about the way nonhuman animals could be unexpectedly elevated by the same economic processes. Novelists critical of political economy were particularly worried by the ways its methods of connecting natural and social worlds could throw the relationship between the human and the nonhuman into disarray.

According to the classical economic doctrine prevalent in the early Victorian era, the best method of sorting and assigning value to the many parties connected through systems of global exchange was the price each commodity or laborer could fetch on the open market. Novelists used their writing to expose the deadening effect that such monetized relations had on the traditional ethical ties that bound people together. But they were equally concerned that the practice of treating price as an accurate measure of value was upending established moral hierarchies between species. The result was an unwieldy fusion of nature and society that neither naturalists nor economists were capable of putting to rights.

In *Mary Barton*, the dehumanizing effects of economic logic find their most obvious, most villainous embodiment in the figure of Harry Carson. Harry, the dashing young heir to his father's manufacturing business, has set his sights on the beautiful but low-born Mary. Mary thinks she loves Harry, despite the fact that their respective classes complicate the relationship. Her vanity is tickled at the prospect of being rich and being loved by someone of Harry's status. Self-interest is not really the foundation of her decision-making, however. Mary genuinely believes she loves Harry, and she is entranced by the idea that wealth could rescue her father from poverty and depression. Her ego is only a minor factor in her flirtation; she is far more invested in dreaming about the things she could do for her loved ones if she were rich. In her fantasy of acquiring wealth, "[e]very one who had shown her kindness in her low estate should then be repaid a hundredfold" (79). For Harry, by contrast, Mary is first and foremost an object of desire. He sees her sexuality as a good or service, and all his flirting is little more than a bargaining process to maximize his personal profit, even if it comes at Mary's expense: "[S]he had the innocence, or the ignorance, to believe his intentions honourable; and he, feeling that at any price he must have her, only that he would obtain her as cheaply as he could, had never undeceived her" (131–32).

The inhumanity of this economic mindset lies in the way it trusts market pricing to determine value. For Harry, Mary is worth exactly what she makes him pay for sex with her—no more, no less. If Mary lets him sleep with her after some flirting and idle promises, he will enjoy the freely acquired pleasures of her body and move on without a further thought of her. If Mary insists on waiting until marriage, Harry will pay the price of the wedding, of monogamy, and of lifelong financial support in exchange for the same pleasures. To Harry, there is no need to anguish over the fact that one of these transactional relationships leads to disrespect and misery for Mary while the other offers her a life of comparative ease and contentment. The moral contradictions between the two futures—the very different values they assign to Mary herself—are suspended. Harry has perfect faith that such contradictions will be resolved by the question of purchase price. How Mary bargains with her body will determine her value; Mary is just another commodity whose worth is equal to and exhausted by the price decided in its exchange.

This belief that the negotiated price of a good decided its inherent value was a common maxim of the classical political economy dominant in the first half of the nineteenth century.[21] Harry's reliance on it demonstrates his own education in contemporary economic theory—and Gaskell's, too. This theory of value was intended to explain the increasingly complex, interconnected world of exchange. It promised to streamline the decision-making process for individuals trying to navigate such entangled worlds to their own advantage. Trusting markets to determine pricing, and trusting pricing to indicate inherent worth, helped reduce the dizzying complexity of such networks to the point where making decisions within them could seem simple, straightforward, and eminently logical. Yet precisely because this logic of valuation was applied within a dynamic, interconnected network, it had unforeseen material and moral consequences. One of those consequences was the kind of callous dehumanization evident in Harry's treatment of Mary. Another real but less obvious consequence was the increased value assigned to economically productive nonhuman beings.

In an unfettered marketplace, the prices of meat, laboring animals, and natural resources can and do surge above the wages of human laborers. Such scrambled species hierarchies are featured prominently in the industrial novels, where they cause angry outcries from working-class characters and the authors who sympathize with them. Philip Warner, the thoughtful weaver in Disraeli's *Sybil*, insists with bitter eloquence that his own degenerate state results from the perverted commercial value system, a system that justifies distributing resources away from laborers and toward business owners and livestock: "The capitalist flourishes, he amasses immense wealth; we sink, lower and lower; lower than the beasts of burthen; for they are fed better than we are, cared for more. And it is just, for according to the present system they are more precious."[22] In Kingsley's *Alton Locke,* Alton takes these confused relations between humans and animals to be the iconic image of a society turned topsy-turvy. The fact that luxurious meat and wool fetch better prices than human labor is typical, he notes, of a disrupted moral order: "I went on, sickened with the contrast between the highly-bred, over-fed, fat, thick-woolled animals, with their troughs of turnips and malt-dust, and their racks of rich-clover hay, and their little pent-house of rock-salt, having nothing to do but to eat and sleep, and eat again, and the little half-starved shivering [human] animals who were their slaves. Man the master of the brutes? Bah! As society is now, the brutes are the masters—the horse, the sheep, the bullock, is the master, and the labourer is their slave."[23]

This animal-centric moral order is a far cry from the merry bartons of preindustrial society. It emphasizes how modern political economy leaves the working classes trapped in a strangely undefined space, caught between the increasingly inadequate categories of nature and society, human and brute beast. But Kingsley's terminology does not portray the British working classes as being completely alone in these categorical borderlands. In describing the farm animal as "the master" and the worker as a "slave" to other creatures, Kingsley yokes the scrambling of class and species hierarchies to the racist hierarchies of plantation slavery. He implies that British workers shared this abominable experience of quasi-humanity with the enslaved Black and Indigenous laborers on plantations in the Americas and elsewhere. It's a complex and problematic comparison, but an instructive one. It helps clarify how ecological plots existed uneasily alongside other moral concepts used to navigate questions of rights and ethics in the Enlightenment tradition. In particular, it foregrounds how nineteenth-century notions of humanity were bound up with debates over slavery, race, and species—and how ecological plots disrupted such categories in unforeseen ways.

Well-known accounts of slavery written by individuals who escaped it often dwelled on the same upended human-animal hierarchies so prominent in Kingsley's novel and other industrial fiction. Alton's complaint, for example, closely echoes passages from the American statesman, orator, and abolitionist Frederick Douglass, who fled his enslavers in 1838. In *Narrative of the Life of Frederick Douglass* (1845), he reflects on the humiliation and horror of being commodified on the auction block alongside a bustling variety of nonhuman farm animals: "We were all ranked together at the valuation. . . . There were horses and men, cattle and women, pigs and children, all holding the same rank in the scale of being. . . . Our fate for life was now to be decided. We had no more voice in that decision than the brutes among whom we were ranked."[24] In his reference to "the scale of being," Douglass invokes the Great Chain of Being, a system that philosophers and theologians had long used to sort the world into a clear hierarchy. It placed God and the angels at the top as the pinnacles of divine perfection. Just beneath them were human beings, caught between their divine and animal natures. Animals fell below humans, with plants and minerals at the base.[25]

Intermixed with Douglass's experience of the monstrous levelling of the auction block were other experiences where slaveholders clearly prioritized nonhuman animals over and above enslaved human beings. Recalling the figure of Colonel Edward Lloyd, the owner of the plantation where he was born, Douglass writes, "[I]n nothing was Colonel Lloyd more particular than in the management of his horses. The slightest inattention to these was unpardonable."[26] Lloyd routinely whipped enslaved men for paying too little attention the welfare of his beloved stable.

As Douglass describes it, the horror of these experiences stemmed from his mistreatment at the hands of a system that did not recognize or respect his humanity. In the *Narrative,* he adopts an Enlightenment understanding of human rights as a unique moral birthright for the species, one springing from the rational nature of humanity itself. His moving account of his mistreatment reasserts that humanity to anyone who might still doubt it and condemns slavery for the way it perverts the natural order by refusing to respect the moral personhood of enslaved people. In this philosophical tradition, the moral status of humanity excludes human beings from the kind of objectification that is routinely applied to other creatures—and in the process it reaffirms a clear division between the human and the nonhuman. Douglass's later speeches would clarify that he abhorred cruelty to nonhuman laboring animals as well. "The master," Douglass lamented in one lecture about the disastrous moral effects of slavery, "blamed the overseer; the overseer the slave, and the slave the horses, oxen, and mules; and violence and brutality fell upon animals as a consequence."[27] Still, the bulk of Douglass's public political arguments rested on a certainty that his audience subscribed to an ideal of human rights that asserted a special and superior moral position for human beings. The chief horror of slavery, in this account, was its tendency to obscure and confuse this self-evident moral taxonomy.[28]

When Kingsley used the terminology of slavery to highlight the vexed relationship between human and animal laborers in *Alton Locke,* then, he was turning to—and effectively appropriating—the experiences of actual enslaved people. It's even possible that Kingsley encountered the words of Douglass and other formerly enslaved activists directly. Given Kingsley's "life-long stand against slavery," the author of *Alton Locke* may have read such testimonies or even heard them aloud as part of the international lecture circuit that brought American abolitionists across the Atlantic.[29] (Douglass himself embarked on an extensive speaking tour of Britain and Ireland for nineteen months, from 1845 to 1847.[30]) Whatever the source of the comparison, it blurs key historical, racial, and economic distinctions between the oppression of overwhelmingly white British workers and the enslavement of people of color around the globe. When Kingsley inaccurately insisted that the commodified labor of the former group was identical to the racist objectification of the latter, however, he was not uniquely guilty of making the association. He was engaging with a much larger Victorian conversation about humanity, labor, slavery, and commodification, a conversation struggling to resolve questions about how workers were valued and how that value related to their moral status as human beings.

By the mid-Victorian stage of this debate, the act of comparing British workers to slaves was a rhetorical commonplace. Perhaps unexpectedly, the comparison first gained popularity among slavery's defenders. British writers who saw little need to agitate for abolition routinely compared British workers to enslaved peoples, using their supposedly parallel situations to argue that the exploitation of workers at home was more horrifying and more urgent than the geographically remote problem of American slavery.[31] Britain had outlawed the slave trade in its territories in 1807, and the empire began the process of emancipating all enslaved people who remained within its dominions during the 1830s.[32] Even those progressive midcentury British thinkers who were proud of such advances toward equal rights often felt comfortable classifying slavery—and the racism that facilitated it—as a lingering foreign concern.[33] They saw worker exploitation, by contrast, as an escalating domestic crisis. Staunch abolitionists fought back against this perspective not by rejecting the worker-slave comparison but by adapting it to their own purposes. British antislavery advocates believed that comparing the more familiar and evident sufferings of the working classes to the sufferings of enslaved peoples could bring the distant issue of slavery closer to the British public. The comparison thus enabled abolitionists to plead for the relief of both groups, using the growing concern for industrial workers to direct attention to the miseries of enslaved laborers abroad.

In reality, neither racism nor the horrors of the slave system were actually remote from the world of Victorian Britain. While historical prejudice is hard to measure, racism persisted and arguably increased in the country over the course of the nineteenth century.[34] Economic relationships are easier to quantify, and they conclusively show that Britain's material prosperity was inseparable from slavery and its spoils. By the time the practice was banned in Britain and its dominions, the profits of the slave trade had utterly transformed British society. Over the course of the 1600s and 1700s, the slave trade secured the nation's prominent role in the international exchange of crops like sugar, tobacco, cotton, and rice. It also catalyzed the growth of London as a financial center for the banking and insurance industries, the sectors that underwrote the entire maritime world. And even after Britain had outlawed slavery, the textile business that enriched the middle classes through the first half of the nineteenth century continued to rely on cheap cotton produced by slave labor in the United States.[35]

Ironically, the habit of comparing exploited British workers to enslaved laborers overseas actually obscured such material connections. The worker-slave comparison treated British capitalism and American slavery as parallel systems, separate economic structures whose lowest rungs were occupied by analogous oppressed groups. But the two systems were not truly operating in parallel at all; they remained closely tied to one another. Banning slavery in British territories was an undeniable moral good, but it did not actually sever the relations created by the flow of capital, goods, and bodies between British industry and American slavery. It simply transformed which kinds of bodies and materials were permitted to cross which geographic boundaries in an ongoing, unbroken international network.

While worker-slave comparisons distracted from these connections, the ecological plot had the power to bring them to light. It narrated such networked relations in new ways that were easily comprehensible to the general public. If the power of such plots to unsettle established geographic and moral taxonomies first came to the fore in Malthus's *Essay*, it was also apparent in Harriet Martineau's writing about a remote Scottish island in "Weal and Woe in Garveloch." There, Martineau's plot emphasized that the health and welfare of the locals depended not just on individual decisions and regional weather patterns but also on the transoceanic migrations of shoals of fish, on the number of fishing stations that competed for the same shoals at various stages in their life cycles, and—tellingly—on "the diet of the West India Slaves."[36] In Martineau's story, then, the neat parallels established by the worker-slave comparison—the idea of two oppressed groups united only by the resemblances between their sufferings—are demolished. In their place is a dawning recognition that impoverished denizens of the British isles and enslaved laborers in America were not distinct, somewhat analogous groups. They were different parts of the same interconnected system, and as such they shared entwined and interrelated fates.

This recognition of material entanglement had the power to open up a new ethical understanding based on interdependence—and new spheres of political possibility as well. Martineau's adherence to economic doctrine may have limited the amplitude of her messaging, but the expansive vision enabled by her ecological plotting still yielded unexpectedly forward-thinking political and ethical insights. Narrating the connection between British consumption habits and their overseas effects in "Cinnamon and Pearls," for example, had led Martineau to paint a damning picture of the European scramble for Sri Lanka's natural resources. (Disappointingly, the story of Martineau's that deals most directly with slavery, "Demerara" [1832], prioritizes dialogues that lay out philosophical arguments about natural rights rather than narratives that show how free British subjects were implicated in the enslavement of others overseas.) The point is not that ecological plots immediately led to a single, incontrovertible politics or ethics explaining how communities should function. After all, recognizing that disparate groups of people, animals, and plants are materially dependent on one another does not resolve the question of how their complex and conflicting needs should be met. Moreover, most of the novelists who employed the ecological plot did not set their novels abroad or extend them across international networks in the ways Martineau had done. They were busy enough on the home front, trying to trace and understand even the most local connections that knit their domestic communities together.

Nevertheless, ecological plotting made it increasingly difficult to deny the existence of such material connections and the interdependence they implied. In the process, these stories raised questions about how, exactly, human beings could be extricated from such networks and assigned an isolated, privileged moral status that set humanity apart. While the idea of horses, sheep, and cattle being prized more than human beings remained outrageous, the response that once seemed both rational and self-evident—the Enlightenment appeal to inherent rights based on the special moral status of humanity—began to look troubled. Through the simple act of tracing how matter and energy moved across the imaginary borders separating nations, races, classes, and species, the ecological plot began to erode those boundaries. In place of the old hierarchical taxonomies of difference, ecological plots offered a glimpse of a world defined by reciprocity, relationality, entanglement, and kinship. They raised the possibility of an ethic rooted in solidarity, dependence, and communal care, forms of mutualism that transcended the categories of humanity and personhood altogether. Such a possibility might seem intriguing or even welcome today.[37] When it arose and began to unsettle Enlightenment notions of the human, however, it could appear confusing, disturbing, and even monstrous.

This monstrosity rears its head in surprisingly literal ways across the industrial novels themselves. These otherwise staunchly realist works feature a disproportionate number of chimeric characters: figures whose weird intimacy with both nature and society marks them as neither animal nor fully human.[38] Some of these figures are real players in the stories themselves, such as the centaur-like circus performers in Dickens's *Hard Times* (1854) or the lawless, animalized miners known as the Wodgate Hell-Cats in Disraeli's *Sybil*. Others are obviously imaginary and figurative. In *Alton Locke,* for example, the workers' frustration at seeing horses and cattle prioritized above humans is not the novel's final word on interspecies relationships. Toward the end of the book, the supposedly self-evident outrage of dehumanization is complicated by a chapter-long fever dream. In it, Alton individually evolves through a series of animal forms (colonial coral polyps, a crab, a remora, an ostrich, a giant ground sloth, an orangutan, and a prehistoric human being) as he struggles with just how difficult it is to explain what, exactly, justifies the tradition of valorizing humanity over the rest of creation.[39] But it is once again *Mary Barton* that provides the most striking illustration of how such hybrid figures work and why they matter so much to the version of ethical interconnection these writers were struggling to define.

In *Mary Barton,* the often unfathomed connections between human society and nonhuman nature surface in the form of a mythical sea creature strikingly out of place in industrial Manchester: a mermaid. The mermaid enters the story shortly after the appearance of Will Wilson, a sailor who promises to entertain Mary, her friend Margaret, and Margaret's grandfather, Job Legh, with tales from his travels—including an amusing yarn about a mermaid he heard secondhand from a friend. As Will describes her, the mermaid is effectively indistinguishable from a human being from the waist up. She clearly breathes air and feels cold like people do. She emerges from the depths "puffing, like . . . folks in th' asthma. . . . [She] had come up to warm herself." She is "as beautiful as any of the wax ladies in the barbers' shops." In keeping with her alluring appearance, she holds a mirror and a comb she employs to brush her long locks. As if to emphasize the fact that she is "just like any other woman," the listeners to Will's story begin mimicking the mermaid in acts of unconscious bodily sympathy.

When the mermaid comes up puffing for air, Mary begins to ask questions "breathlessly," and Job Legh starts "to smoke with very audible puffs." The mermaid in Will's story never speaks, but she emphasizes the reflections between herself and her fascinated human onlookers through literal acts of mirroring, sometimes turning her looking glass upon herself and other times "holding up her glass for [the sailors] to look in" (147).

For all the mermaid's apparent humanity, however, her admirers can only conceive of her value in commodified form. When the sailors first saw her, Will explains, "They all thought she was a fair prize, and may be as good as a whale in ready money (they were whale-fishers, you know)" (147). Even Mary, who closely resembles the mermaid in several respects, seems unable to tap into the sympathy that would cause her to question such economic logic. Like Mary, the mermaid is a beautiful woman who finds herself pursued by multiple men. Like Mary, the mermaid sometimes encourages the attentions of these men and sometimes spurns them—a telling sign, from Will's perspective, that "one half of [the mermaid] was woman" (148). And like Mary, the mermaid is thoughtlessly commodified by the men she entices. Nevertheless, Mary's desire to know more of the mermaid can be expressed only in acquisitive terms, and she joins the rest of the group in lamenting the mermaid's freedom: "'I wish they had caught her,' said Mary, musing" (148). The infectious power of economic thinking is on full display here, as the most exploited and dehumanized segments of this industrial society nevertheless fall into the trap of responding to others in an analogous position with the same exploitative, dehumanizing mindset.

What makes the mermaid such a potent symbol in *Mary Barton*, however, is the way she manages to represent two distinct consequences of nineteenth-century economic thought at the same time. Her semi-human status leaves her just human enough to exploit without raising questions of ethics. In that respect she serves as a clear parallel to the industrial laborers who discuss her, an entire class of people who find their suffering dismissed by capitalists who characterize them as "more like wild beasts than human beings" (177).[40] At the same time, the anatomical configuration of the mermaid's body conjures up another kind of intimacy between the human and the nonhuman. The mermaid consists of a human portion tied to and dependent on a mass of animal matter that is typically hidden from view. She is, in short, a fitting symbol of the veiled reality of interspecies interdependence, the interdependence that Malthusian thought had recently hauled up for the inspection of the general public.

The instant objectification of the mermaid at the hands of those around her thus offers a damning critique of the prevailing economic mindset. Her treatment suggests that economic thought was incapable of assessing the value of marginalized and dehumanized groups relegated to the outskirts of society, be they mermaids or mill workers or enslaved laborers. But Gaskell's scene also suggests that the newly evident, almost fantastical interdependence of humanity and the nonhuman was materially bound up in this failure. The unbelievable oddity of *Mary Barton*'s mermaid puts the shock of interspecies interdependence on full display. In the process, it suggests that questions of human value were intimately linked to questions of humans' material dependence on the nonhuman world. The difficulty of parsing these relationships could not be left to philosophical assertions of human uniqueness or to mindless market mechanisms. Some other, better way of navigating interspecies relationships was necessary.

EMPIRE, TRADE, AND THE TEMPTATIONS OF NATURAL HISTORY

As it happened, there was another approach to studying interconnection that was wildly popular at the time: natural history. For the first half of the nineteenth century, natural history served as the umbrella term for all forms of study that took the nonhuman world as their focus, from a lady's amateur fern collecting to Charles Lyell's expert attempts to discern the age of the earth through the analysis of geological strata.[41] *Mary Barton* appeared more than a decade before the field would be revolutionized by Darwin's application of Malthusian ideas to natural history. Nevertheless, Darwin's name was well known. The young naturalist's journals detailing his experiences, observations, and collections during his voyage on the *Beagle* had been a surprising success, even if he was known better as a writer than as a scientific thinker. It is this figure of the naturalist—as a collector and trader whose work closely corresponds to the rise of empire and the global marketplace—that crops up in Gaskell's novel. Years before natural history was upended by Malthusian ideas, then, *Mary Barton* was already raising concerns that natural history's entanglements with political economy made it difficult for the field to resolve the emerging ethical dilemmas of interconnection.

The field of natural history is represented in *Mary Barton* through the lovable but divisive figure of Margaret's grandfather, Job Legh. Strictly speaking, Job makes his paltry living as a weaver; he is solidly working-class. But he is also an avid and experienced naturalist. As unique as that combination may sound, the lack of scientific professionalization at the time made such working-class naturalists far from unusual.[42] By the time readers meet him, the cantankerous Job has amassed a collection of specimens from around the world—and the elaborate vocabulary and knowledge to describe them, too. In fact, when the mermaid first enters *Mary Barton* as part of a larger conversation about natural curiosities, Will Wilson brings up the subject partly to capture Job's attention. Surely if anyone were capable of putting the confused relations of humans and nonhumans to rights it would be such a naturalist, a scholar who could survey the breathtaking variety of living things and arrange them into neat logical categories. Yet *Mary Barton* and the other industrial novels suggest that naturalists were just as ineffectual as political economists when it came to bringing order to the chaos of relations—in part because the two groups were already engaged in a messily symbiotic relationship of their own.

In many respects Job is an admirable and even beloved figure. His learning is eccentric but impressive, and his love for Margaret and the other members of his working-class community is undeniable despite his crusty exterior.[43] His good heart becomes clear in the second half of the novel, when he pulls strings to secure a competent lawyer to defend Jem Wilson at the murder trial. It is a task Job can accomplish, despite his low social status, because his years as a naturalist have connected him with a wide range of middle-class professionals in the region who share his idiosyncratic interests.[44] But the fact that Job is so invested in these networks of exchange also has less savory consequences. It transforms him into a narrowly self-interested economic individual, the kind of *Homo economicus* whose shallow ethics show up more clearly in the character of Harry Carson. Harry's lust for Mary leads him to overlook the kinds of moral questions that should structure interpersonal relationships. In Job's case, it is an unshakable lust for specimens that causes him to lose sight of his duties to those around him. Despite his good intentions, Job is so fixated on trading, acquiring, and purchasing creatures for his collection that he fails to acknowledge the value of other lives—a failure that contaminates his relations to others and his ability to prioritize collective well-being.

The novel's resident naturalist is never explicitly presented as a political character. Whenever Job has an opportunity to insert his commentary on the novel's central events, however, his free-market ideology sets him apart from those around him. When John Barton is chosen as a delegate to carry the People's Charter to Parliament, for example, many members of the community approach John with messages to deliver to the government. Job stands out from the pack for his bewildering insistence that what the poor need is more unfettered capitalism. "Yo take my advice, John Barton, and ask Parliament to set trade free, so as workmen can earn a decent wage," he entreats (86). Later, as John crashes around his quarters upstairs plotting revenge on the factory owners, Job sits below with Mary airing his grievances about unionized labor. The naturalist bitterly admits to paying union dues, but not by choice. "I were obliged to become a member for peace, else I don't agree with 'em," he explains. His disagreement is based on economic principles and the individualism that undergirds them: "[N]ow that's not British liberty, I say" (191–92). Even when describing his relation to organized labor, he cannot help couching his account in entrepreneurial metaphors. As he describes it, he is "a sleeping partner in the [union's] concern" (191).

Ironically, the naturalist's economic certitude does not translate into sound financial management. Job literally fails to account for what he owes to those closest to him. He mishandles his household income precisely because he is so narrowly focused on gratifying his self-interest, prioritizing his commodified creatures over all social relations, including those to his own family. "Mary," Margaret confesses tearfully toward the beginning of the story, "we've sometimes little enough to go upon. . . . For grandfather takes a day here, and a day there, for botanising or going after insects, and he'll think little enough of four or five shillings for a specimen; *dear* grandfather!" (47; emphasis mine). "Dear" turns out to be an epithet attached to Job Legh more than once. It resurfaces again in the final pages of the novel, after Jem and Mary emigrate to Canada. As they discuss news from Manchester, they are surprised to learn that Job is seriously considering a transatlantic trip to visit them. As Jem explains,

> ". . . Job Legh talks of coming too—not to see you, Mary,—nor you, mother,—nor you, my little hero" (kissing [his son]), "but to try and pick up a few specimens of Canadian insects, Will says. All the compliment is to the earwigs, you see, mother!"
>
> "Dear Job Legh!" said Mary, softly and seriously. (379)

Safe in their suburban cottage, surrounded by the literal fruits of Jem's success, Jem and Mary are free to laugh at the way Job's nature studies have led him to prioritize insects over people. At second glance, however, they are not actually laughing. Mary offers Job a mixed benediction, delivered "softly and seriously." Job, she declares once again, is "[d]ear," a word with very different emotional and economic meanings. Emotionally speaking, to be dear is to be cherished or held valuable. Job has certainly proven valuable, both in his commitment to rearing his orphaned granddaughter and to helping Jem's legal defense. "Dear" has another related meaning, though. It describes an object that is exceedingly costly, beyond the reach of what one is willing or able to pay. This definition applies to Job as well as the other. He is beloved but costly, an eccentric character whose fascination with nature causes him to grossly overvalue some things and undervalue others. He is, in short, both an asset and a liability to anyone trying to navigate the complex ethical connections of the industrialized world.

Job's inability to see how his habits affect those around him is not simply the isolated error of one obsessive naturalist. His oversights spring from the same issue that concerned Malthus in the *Essay:* the tendency of the purchase price of objects to obscure the all-important question of whether their circulation aids or impedes the flourishing of the community. When Malthus connected the starvation of the poor to an aristocratic love of beef and horseflesh, he cut through the opaque mediation of pricing to highlight the sorts of trade-offs we make every time we accept price as the representation of value in an interdependent community. In the figure of Job Legh, Gaskell suggests that both of the emerging forms of expertise devoted to explaining such networks of relations—natural history and political economy—fail to practice even Malthus's rudimentary ethical accounting. If anything, experts trained in these discourses become so invested in their own metrics that they are actually worse than amateurs at accurately valuing the world that surrounds them. Moreover, Gaskell worries about how closely the two schools of thought are interlinked. Naturalists, she suggests, founded their expertise on exploitative international trade networks, and they indulged in the kind of self-interested commodification of living things sanctioned by the doctrines of political economy.

Mary Barton sounds an early alarm about the problematic consequences of this intellectual traffic, but its ethical perspective was echoed by other practitioners of the emerging social-problem genre. The same mutually reinforcing relationship between political economy and natural history can be seen, for example, in the figure of Thomas Gradgrind, the school superintendent in Dickens's *Hard Times*. Gradgrind is an obvious send-up of the economic mindset: he names his children after Malthus, James Mill, and Adam Smith, and he insists on the importance of facts and the worthlessness of sentiment and imagination. After the publication of *Hard Times*, the word *Gradgrind* actually entered the popular lexicon as a name for a cold, unfeeling person who values quantitative logic alone.[45] In the classroom, however, Gradgrind's first prompt to his pupils is drawn not from political economy but from natural history. "Give me your definition of a horse," he asks one of the main characters, whom he recognizes only as "Girl number twenty."[46] The prompt itself suggests the close alliance between political economy and natural history. Girl number twenty, better known as Sissy Jupe, finds the demand baffling, despite her deep familial intimacy with horses. (Her father is a horse rider at the circus.) When she falters, Gradgrind turns his attentions to the perfect student of political economy, a pallid and heartless boy named Bitzer. True to form, Bitzer provides a technically perfect answer that is, from a practical perspective, all but useless: "Quadruped. Gramnivorous. Forty teeth, namely twenty-four grinders, four eye-teeth, and twelve incisive. Sheds coat in the spring; in marshy countries, sheds hoofs, too. Hoofs hard, but requiring to be shod with iron. Age known by marks in the mouth."[47] This long-winded catalogue of features that would not help anyone recognize a horse, much less understand the animal's vital role in Victorian society, follows *Mary Barton* in suggesting the worthlessness of both natural history and political economy to clarifying the lived relationships binding humans to the natural world.

There is compelling evidence from naturalists themselves that their fascination was driven by the same kind of self-interested acquisitiveness that worried social-problem novelists. "The single-mindedness that [naturalists] gave to their studies could at times be almost inhuman," writes the historian David Elliston Allen. He singles out the example of Philip Henry Gosse, whose journal entry on the day his only child was born reflects the unusual priorities of the obsessive naturalist: "Received green swallow from Jamaica. E delivered of a son."[48] Alfred Russel Wallace, the naturalist and biogeographer, provides even more unsettling examples of such skewed perspectives in his published work.

Wallace is generally co-credited alongside Darwin with developing the theory of natural selection—a feat each achieved on their own after reading Malthus's work. But Wallace lacked Darwin's independent means. He made his money from specimen collecting, spending years abroad in search of animals to kill, preserve, and sell to naturalists back in Europe. Wallace's financial insecurity left him keenly aware of the horrors of economic inequality, and he sometimes felt compelled to comment on such issues in his work. Thus at the end of his popular account of his travels in South Asia, *The Malay Archipelago* (1869), Wallace turned rather abruptly to the topic of social injustice in Britain. Weighing the state of British society against his observations of tribal communities widely denigrated as examples of "the savage state," the naturalist decided that the tribal communities were ethically superior.[49] Britain's scientific and economic advancement had led to a kind of callousness toward the poor that Wallace saw as a moral atrocity. "[A]s regards true social science," he concluded, "we are still in a state of barbarism."[50] Yet in these most active phases of his collecting career, Wallace managed to erect a sort of mental barrier between his humane moral stances and his scientific fieldwork. Later in life he would begin to integrate the two, warning of the dangers of disrupting ecosystems and eventually emerging as an outspoken socialist.[51] In *The Malay Archipelago*, however, natural history's habit of detaching ethics from the study of interspecies relationships is on full display.

Wallace's success as a collector sprang from his ability to understand the chains of narrative cause and effect that connected human and nonhuman populations. In his book, he regularly constructs his own ecological plots to explain the distribution of valuable and unique specimens. So, for example, his unusual success collecting beetles, butterflies, and wasps around a new coal mine in Borneo leads him to realize that "[t]he quantity and the variety of beetles and of many other insects that can be collected at a given time in any tropical locality, will depend, first upon the immediate vicinity of a great extent of virgin forest, and secondly upon the quantity of trees that for some months past have been, and which are still being cut down, and left to dry and decay upon the ground."[52] His insect haul at one particular location is the upshot of a specific set of interactions between human doings and insect preferences: "For several months from twenty to fifty [Chinese and Iban workers] were employed almost exclusively in clearing a large space in the forest, and in making a wide opening for a railroad.... [S]awpits were established at various points in the jungle, and large trees were felled to be cut up into beams and planks. For hundreds of miles in every direction a magnificent forest extended ... and I arrived at the spot just as the rains began to diminish and the daily sunshine to increase.... The number of openings and sunny places and of pathways, were also an attraction to wasps and butterflies."[53]

For all their ecological import, however, Wallace's plots explaining the interactions among human, plant, and animal communities do not lead him to any sudden ethical awakening. His attention to these relations, like Job Legh's, is exclusively a matter of intellectual and commercial interest. That much becomes clear when Wallace switches his quarry from insects to a far more human animal, the orangutan. Wallace makes no moral distinction between insects and anthropomorphic apes, mercilessly seeking out and shooting every orangutan he can find. Many of the early pages of *The Malay Archipelago* are devoted to detailed descriptions of the orangutans "groaning and panting" in pain as they heroically attempt to swing, climb, or run to safety.[54] Nevertheless Wallace continues to pursue them, shattering their arms, legs, hips, spines, and jaws with volley after volley of gunfire until they finally succumb. His deadened moral sense is especially striking when he uncovers "a young [orangutan] face downwards in the bog." He quickly realizes that the adult female he has just killed must have been its mother. "Luckily," he writes, "it did not appear to have been wounded, and after we had cleaned the mud out of its mouth it began to cry out, and seemed quite strong and active."[55]

The episodes that follow are oddly touching. Wallace proceeds to bring the baby home and coddle it like his own child. He bathes it, brushes its hair, makes toys for it, feeds it castor oil for an upset stomach, and carries it with him on his daily rounds. Sadly, his unexpected pet dies of fever three months later—at which point Wallace promptly skins it, boils the flesh from its bones, and prepares its skeletal specimen for shipment to England.[56] Throughout this parenting experiment Wallace continued to shoot other orangutans, which predictably ascended high into the trees once injured. After he killed them, Wallace had to decide whether to abandon their bodies to rot or to retrieve them. Eminently logical, Wallace based his reasoning on careful pecuniary calculations about how much each specimen would likely fetch on the market—and how his payments to local workers to retrieve any given body might affect their wage demands for future retrievals.[57]

There is a consistent logic, and even admirable equality, to this vision of a networked world. It is of a nightmarish kind, however, as it valorizes a mindset that calculates everything in terms of material gain rather than moral significance. For some, that was exactly the appeal: political economy promised to help individuals navigate the intricacies of interconnection by setting aside questions of moral priority and replacing them with the simpler matters of market price and calculated self-interest. It was a worldview that leveled distinctions and brought humans, plants, and animals closer together—but only so they could all be exploited equally by those whose monetary interests automatically trumped ethical hesitations. The naturalists who might have imposed some sort of order on this flattened worldview were often so entrenched in it themselves, and so busy cataloguing and commodifying the species they studied, that they offered little help.[58] The social-problem novelists stepped in to return moral questions to the heart of the discussion. They accepted the idea of interconnection between classes and species that economics unveiled. When it came to ethics, however, they strove for a better way of ascertaining moral value.

TOWARD A RENEWED FAITH IN SOCIETY

As the social-problem novelists soon discovered, there is a big difference between identifying an issue and providing a convincing solution to it. The hearts of social-problem novelists were undeniably in the right place, but even admirers of the genre tend to agree that these fictional projects are riddled with shortcomings. Over time, the early Victorian works have become notorious for their failure to offer practical, satisfying cures for the maladies they diagnosed. Their politics haven't aged well, either. While their expressions of sympathy for the working classes might have looked risky and even radical in an era of widespread political instability, the novels' middle-class authors shied away from concrete suggestions for reform (much less overt calls for revolution).[59] What these writers urged instead was a deeper feeling for and identification with the interests of the community, the "sympathetic 'we'" of Gaskell's boy down by the docks. The result was a sort of moral compromise between opposing sides, one that critics have long dismissed as morally concerned but effectively apolitical.

Yet even this apparently restrained and noncommittal position unearthed some surprisingly radical possibilities. Emphasizing the shared ethical interests of the community and its expansive nature raised the question of where, exactly, the outer boundaries of a moral community ought to be drawn. In a world of potentially endless connection and interdependence, who counted as ethically significant and who didn't? And given that so many seemingly distinct individuals, classes, and species were connected by networks of exchange that remained largely unmapped, how might one seemingly innocuous decision in one place have ethically unsettling impacts elsewhere? Political economy and natural history had punted on these moral quandaries. The social-problem novelists were at least determined to acknowledge and investigate them.

The new confusion about upended natural and moral orders appears in these works in a variety of ways. It registers in the complaints of working-class characters who feel themselves implicitly ranked beneath domesticated animals—and in their employers, who curse the workers as worse than beasts. It surfaces in the strange proliferation of personified animal characters scattered across the pages of these works, from Harold the noble bloodhound in *Sybil* to the mutt Merrylegs who symbolizes the mystery of love in *Hard Times*. It crops up in the form of human-animal hybrids who become centers of discussion in *Mary Barton* and *Alton Locke*. In every instance these texts are plagued by a proliferation of marginal cases, of specimens and characters whose exact relation to the political and ethical protections of society is unclear. These are novels self-consciously vexed by the question of how far society and its protections should extend.

In short, the social-problem novels are haunted by the problem of the social itself. While their most explicit concern was the inequality brought about by industrialization and its economic doctrines, their philosophical concerns ran much deeper. They recognized that the nature and definition of community had been transformed by the interconnections that political economy had revealed. Class and species barriers no longer looked like self-evident ways of defining a community or assigning priority within it. Collective interests stretched beyond these boundaries, passing along lines of interdependence that remained mostly unmapped. Materially speaking, neither the social nor the human seemed like isolated and easily defensible categories anymore. Yet there was no obvious substitute for them, no clear candidate to replace the older ways of thinking that would answer the moral questions raised by material interdependence.

Faced with the intractable problem of imposing limits and assigning priorities in a theoretically endless network, the social-problem novelists sought solutions outside the material realm. They have often been critiqued for this, too—accused by political critics of retreating from actual worldly struggles to seek refuge in watered-down forms of Christian fellowship. *Mary Barton*'s climactic reconciliation between masters and men, for example, depends on an impromptu all-night Bible study (356–57). The industrialist Carson emerges from his scriptural readings transformed, ready to forgive his son's killers and reopen negotiations with laborers to

ensure that "the Spirit of Christ," rather than economic doctrine, serves "as the regulating law between both parties" (374). Charles Kingsley, who was not only a novelist but also a naturalist and an ordained clergyman, is even more explicit on this head. A lengthy speech toward the end of *Alton Locke* emphasizes that human rights are a political dead end. The very definition of the human, he points out, is based on a smattering of anatomical characteristics that are subject to renegotiation based on the interests and whims of those in power. He particularly decries the way human diversity has repeatedly been used to police the boundaries of humanity and deny full moral standing to people of color. "Claim your investiture as free men from none but God," one saintly character proclaims. "His will, His love, is a stronger ground, surely, than abstract rights and ethnological opinions. . . . Looked at apart from Him, each race, each individual of mankind, stands separate and alone, owing no more brotherhood to each other than wolf to wolf, or pike to pike."[60]

It is easy to dismiss these sudden pivots to religion as evasive. They offer examples of a very literal kind of deus ex machina, as novelists try to impose tidy conclusions on the conflicts of mechanized modernity through divine intervention. But the fact that industrial novelists found it necessary to turn away from the material plane to resolve the problems that emerge from material interdependence is not really a weakness. It shows that these writers were aware that interdependence scuttled easy claims to moral priority. Humans and the societies they formed were no longer strictly separable from nonhumans and the natural world; they were bound to each other through complex and poorly understood webs of reciprocal relations. There was no simple way to define humans as distinct anymore. There was also no way to be sure how a local decision in one of these interdependent communities might ripple out to affect its other members. There wasn't even a sure way of knowing in advance who all those community members might be.

The social-problem novelists told stories to illustrate such webs and explore the compounding moral consequences of actions that propagated across them. It should come as no surprise that they had no conclusive answers for how to live ethically in such an entangled world. These are questions we still struggle with almost two centuries later. What the social-problem novelists contributed to such debates was a healthy dose of skepticism, a deep suspicion of all simplifying, self-interested prescriptions of the kind that conclude Martineau's *Illustrations of Political Economy*. Unsettled by the strange inversions of this networked world and at a loss for answers, the social-problem novelists fell back on a different form of moral certainty—the kind offered by faith in a higher power. A divine intelligence might sort out what they could not. For the mostly Anglican and Unitarian authors writing about the Condition of England, Christianity neatly defined the moral community as consisting of human beings created in God's image. And whatever schisms might separate one Christian denomination from another, they all at least shared a fundamental belief in the virtue of loving one's neighbor, of recognizing self-sacrifice as a higher ideal than self-interest.

The religiosity of these works was, in short, a strategic solution to a set of intractable ethical problems. The industrial novelists were among the first modern thinkers to see these problems and take them seriously. They wrestled with the ways interdependence created moral obligations that could upend established species hierarchies. In essence, these writers inadvertently stumbled on the need for a new, more ecological ethic. Their ability to glimpse that ethical horizon set them apart from other contemporary thinkers about interconnection—particularly those in political economy and natural history. The prospect was so unsettling to established ideas of human priority, however, that these writers retreated almost immediately. They quickly withdrew and took shelter in transcendent understandings of humanity grounded in faith, a faith they could safely assume was shared by most of their contemporary readers.

Yet it was a faith that was already eroding. Before long, the Victorian era would become notorious for a general crisis of belief. That crisis arose in part from newly powerful material explanations of the world that sprang from natural history, the conclusions of naturalists like Lyell and Darwin that ran counter to religious assurances about the age of the earth and nature's fundamentally moral structure. Doubts also arose as a result of newly rigorous studies of scripture, particularly those associated with German higher criticism. This approach to biblical scholarship pioneered by Friedrich Schleiermacher, David Friedrich Strauss, and Ludwig Feuerbach investigated the historical production of scripture only to conclude that many of the factual claims in the Bible clashed with the chronological record.[61]

As the century wore on, a moral approach to community rooted in shared faith became increasingly untenable. More skeptical social novelists would need to develop other ways of navigating the ethical complexities of interconnection. They would need to do a better job of clarifying the novel's role in relation to the increasingly ascendant social and natural sciences, too. These writers' secular artistic methods would have more lasting effects than those of the industrial novelists, decisively shaping both the standard conventions of the novel and the techniques modern thinkers would use to imagine humanity's increasingly undeniable dependence on the environment around us.

CHAPTER 3

The Nature of Fiction

Before Marian Evans could become a successful novelist, she had to get her feet wet. In 1856 Evans, the editor of the *Westminster Review,* managed to escape the demands of her job in London to spend several months of summer holiday at the Devon coast with her lover, George Henry Lewes. Together they passed their days writing, visiting local zoologists, and wading through tide pools in search of anemones for their home aquarium. It all sounds rather idyllic. In reality, the trip was a mix of business and pleasure. The couple's rambles along the shore surrounding Ilfracombe served as preliminary research for Lewes's writings on natural history, which culminated in a series of articles he eventually collected and published as *Seaside Studies* (1858).[1]

While her partner pursued natural history writing, Evans took the opportunity to reflect carefully on her own authorial undertakings. Previously, she had stuck to nonfiction and translation. By the end of the trip, however, Evans began to think of herself as a budding novelist. "I am anxious to begin my fiction writing," she confessed in her journal on July 20, 1856.[2] While in Devon she wrote a review that laid out her personal sense of what realist novels should do. She was especially preoccupied with the question of what social fiction could tell us in a world that already seemed exhaustively documented by those other methods of studying interconnection: natural history and political economy. Never one for idleness, Evans began putting her theory into practice almost immediately. Six months later her first short fiction appeared anonymously in *Blackwood's Edinburgh Magazine.* Within a year she was in the thick of writing her first novel, *Adam Bede* (1859). By that time she had settled on the pen name that she made famous: George Eliot.[3]

The early years of Eliot's fiction career provide an exceptional opportunity to see how one of the era's most discerning thinkers understood the evolving relations among political economy, natural history, and the social novel. Eliot is unusual not only for the depth of her insight but also for the meticulous records she kept of her own reading. Her published reviews and private journal entries make it possible to locate intellectual encounters and lines of influence that can only be inferred in the works of most authors. Those records show traffic between economic thought and social fiction dwindling by midcentury, as economic thinkers excised most of the concrete storytelling from their writing to reframe themselves as mathematical theorists pursuing increasingly abstract formulas to model social dynamics. It was during the process of casting about for more concrete, more accessible models of interconnection that Eliot found herself drawn to natural history. A few years after she began writing fiction, however, Eliot's adulation of natural history would also wane. Her growing confidence as a novelist corresponded to a growing sense that the social novel ought to stand apart from other studies of interconnection. The novel, she decided, had a particular job to do. The work of fiction played a complementary role to these other fields, a role that provided the social novel with its own unique moral and artistic mission.

Eliot's progression from amateur naturalist to professional novelist thus provides a telling case study in the social novel's increasing differentiation from both natural history and political economy as the nineteenth century wore on. What makes her trajectory especially significant is her attunement to the ways literary form could help distinguish between these fields. As Eliot saw it, the narrative strategies of delineating characters, placing them in particularized settings, and telling the story of their interactions came together to produce a powerful tool for steering readers' ethical attentions and molding their sense of community. But these were exactly the strategies that midcentury economic thought had abandoned, sacrificing storytelling and particularity in its drive toward mathematical abstraction. In her early meditations on novel writing, Eliot rejected this arid theorizing of community, embracing the messier material practices of natural history instead. She admired naturalists' close attention to the specific life histories of animals and plants, and she saw their studies as a possible model for fiction writing. Her first novel would put that theory into practice. But like the industrial novels that preceded it, *Adam Bede* would stumble over the problems that nonhuman species posed to a clearly defined moral vision. Eliot's second novel, *The Mill on the Floss* (1860), found a way to solve the problems posed by our fellow creatures—not through a transcendent, religious vision of humanity (as in earlier realist works) but through a discerning application of literary form. *The Mill on the Floss* uses literary form to demote other species from character status to the realm of mere setting, treating them as elements of the environmental surroundings set apart from the more central, more involving human drama.

It is worth lingering over Eliot's careful formal separation of human character from nonhuman setting. For one thing, her approach helps clarify how the literary categories of character and setting work in practice. It also offers a window onto the surprisingly contingent historical process that led character and setting to map neatly onto the philosophical distinction between the human and the nonhuman—an equivalence that would have substantial impact on our sense of how other creatures ought to be treated. In the end, much of our ethics and politics depends on what we pay attention to. Eliot's influential later works—including *Middlemarch* (1871–72), an enduring favorite to this day—persistently direct readers away from nature. They use the distinction between character and setting to effectively undo the connections between humanity and the nonhuman world that had been the hallmark of earlier nineteenth-century ecological plotting.[4]

But this is not a condemnation of Eliot's decision to sever ecological ties in her fiction. Although she reimposed a kind of literary separation between humanity and the natural world, Eliot never bought into the idea that such a separation was accurate, or obvious, or inevitable. Throughout her life she treated attempts to isolate humanity from other creatures as an aesthetic and ethical tactic, one that was ultimately arbitrary. She developed sophisticated and enduring literary techniques for highlighting human concerns at the expense of other creatures, but she remained aware of the doubtful moral grounds for doing so. Her career thus provides an ideal place to explore how the techniques we associate with realist fiction privilege the human, why they might do so, and how—under other conditions—they might have developed otherwise.

GEORGE ELIOT AND THE CASE FOR NATURAL HISTORY

Eliot had a very successful career writing and thinking about books long before she penned any of her own. In 1851 she joined the publisher John Chapman in his project of reviving the *Westminster Review*, a journal she helped edit until she turned to fiction writing full-time five years later. Eliot's duties at the *Review* included managing the magazine's stable of impressive writers, composing essays on social and historical subjects, and sometimes penning the anonymous "Belles Lettres" column surveying recent literary output to identify strengths, weaknesses, and general trends in publishing.[5] The extraordinary breadth of Eliot's reading soon led her to the same question that preoccupied so many of the midcentury social-problem novelists: namely, what fictional portrayals of interconnection could possibly add to the increasingly dominant model of society proposed by political economy.

Nevertheless, there were certain glaring differences between Eliot and many of the writers who came before her. While she stands out today as a towering figure in nineteenth-century letters—in some circles her name has become almost synonymous with the Victorian era—Eliot was a renegade by the standards of her day. Her forays into publishing began with her own scandalous translations of both David Strauss's *The Life of Jesus, Critically Examined* (1835–36; trans. 1846) and Ludwig Feuerbach's *The Essence of Christianity* (1841; trans. 1854), two works of German scholarship that cast doubt on the divine truth of the Bible in favor of historical and philosophical understandings of Jesus and his teachings. Her personal life was even more shocking. When she took up her position as clandestine editor of the *Westminster Review*, she moved into the townhouse Chapman shared with his wife and mistress at 142 Strand. During her first stay there she had a romantic dalliance with Chapman. By the mid-1850s she had taken up with Lewes, another married man whose home life was notorious for its infidelity. Although both Eliot and Lewes considered their partnership a marriage, it remained legally invalid, tarnishing Eliot's public reputation and straining her ties with her family.[6]

By Victorian standards, then, Eliot was decidedly radical and nonconformist. Compared to her predecessors, she was far more receptive to the latest intellectual developments of the age, including the rise of economic thought. She was also far less inclined to fall back on Christian faith when faced with the contradictions that arose from these new approaches to community. When she sat down to write about realist novels at Ilfracombe, both these habits of mind came into play. Together, they would shape her sense of what the social novel was and what it should do, leaving a lasting mark on modern realist fiction.

The article Eliot wrote in Devon was among the last reviews she produced for Chapman. Strictly speaking, it was supposed to treat the work of German journalist and social analyst Wilhelm Heinrich Riehl. But the piece, which appeared as "The Natural History of German Life" (1856), ended up tackling much bigger questions. In it, Eliot sought to cut any remaining ties between the novel and political economy, emphasizing the lines of kinship between novelists and naturalists instead. To Eliot the value of Riehl's work lay in its dismissal of economic methods and its embrace of natural history. While Riehl's studies focused on human groups, he treated those groups as subject to the same kind of attentive description that defined so much of contemporary nature study. Eliot read his work and composed her review between forays into Devon's rocky intertidal zone, where she and Lewes collected samples of marine life. As she read and explored, she found herself captivated by the analogies between the coastal communities she sought to re-create in her home aquarium and the human communities Riehl sought to re-create in his writings. Her review praises Riehl for offering a "Natural History of social bodies," one devoted to the particular dynamics of a specific group of interconnected human beings.[7]

Particularity and specificity were precisely what was missing from the increasingly professionalized science of political economy. Political economists studied interconnection in theory only: in their rush to devise abstract, universal formulas explaining trade, economists had lost touch with the specific people and real-world communities for whom such trade actually mattered. Eliot decried "[t]he tendency created by the splendid conquests of modern generalization, to believe that all social questions are merged in economical science, and that the relations of men to their neighbours may be settled by algebraic equations."[8] As she saw it, this drive to abstraction was liable to end in catastrophe. Political economists were overhasty in their attempts to impose a single, streamlined template of proper communal structure onto the messy realities of individual communities and their historically engrained relationships. Careful observation of such communities was an important first step toward cultivating sympathies within and between them. Those sympathies, in turn, were necessary to the judicious practice of public policy.

The evolution of economic thought over the nineteenth century furnishes plentiful evidence of the abstraction Eliot found so off-putting. Even the delightfully essayistic Malthus grew dry and mathematical after he left his ecclesiastical work to join the official ranks of political economists. As the economic historian Lionel Robbins observes, the second edition of the *Essay on the Principle of Population* (1803) "was virtually a new book. The first essay was a brilliant tract from a man just down from Cambridge. The second edition is a rather dull treatise with all sorts of statistics and empirical facts."[9]

Even so, Malthus maintained to the end of his life that political economy could never be exclusively mathematical. "[W]hether we advert to the qualities of man, or of the earth he is destined to cultivate," he observed in his *Principles of Political Economy* (1820), "we shall be compelled to acknowledge, that the science of political economy bears a nearer resemblance to the science of morals and politics than to that of mathematics."[10]

Here, as so often in his life, Malthus was an outlier. Other nineteenth-century advocates of the field were certain that political economy offered dependable laws and definitive mathematical formulations of human behavior. Later, this tendency to simplify human societies into elegant but erroneous and unfalsifiable mathematical models would be dubbed the "Ricardian vice," a name that honors (or really dishonors) Malthus's friend and rival David Ricardo.[11] The economic habit of looking past real-world examples in the search for laws and lessons is clear in the tension between Martineau's complex *Illustrations* and the simplistic economic conclusions appended to them. It is also stated quite nakedly in textbooks of the time, which demonstrate an almost Newtonian obsession with converting lived phenomena into universal laws. "[F]our inquiries are comprehended in this science," wrote James Mill in his *Elements of Political Economy* (1821):

1st. What are the laws, which regulate the production of commodities :
2dly. What are the laws, according to which the commodities, produced by the labour of the community, are distributed :
3dly. What are the laws, according to which commodities are exchanged for one another :
4thly. What are the laws, which regulate consumption.[12]

From the very beginning, this method met with substantial pushback from those who saw such a calculating approach as immoral, unrealistic, and inhuman. William Stanley Jevons was still responding to those objections when he opened his revolutionary *Theory of Political Economy* (1871) with an explicit defense of the abstract, quantitative nature of the field, which by that time was adopting the name it takes today: economics. "It is clear that Economics, if it is to be a science at all, must be a mathematical science," he wrote. "There exists much prejudice against attempts to introduce the methods and language of mathematics into any branch of the moral sciences. Many persons seem to think that the physical sciences form the proper sphere of mathematical method, and that the moral sciences demand some other method,—I know not what. My theory of Economics, however, is purely mathematical in character."[13] While much has changed in economics over the past 150 years, the scientific aspirations of the field remain a driving force in both its form and its development. Since the mid-nineteenth century, economists have understood themselves as professionals who have risen above storytelling, taking pride in their proficiency at depersonalizing social relations in order to translate them into general equations, tables, and quantitative models.[14]

This tension between abstraction and specificity in moral questions was what concerned Eliot at Ilfracombe. As she mulled over the place of the novel in exploring communal issues, she decided that "[t]he greatest benefit we owe to the artist, whether painter, poet, or novelist, is the extension of our sympathies." Such sympathies help bring society together for the good of the whole, "linking the higher classes with the lower . . . [and] obliterating the vulgarity of exclusiveness." But this communal banding-together could not be achieved through the simple deployment of "generalizations and statistics[, which] require a sympathy ready-made, a moral sentiment already in activity."[15] Social solidarity must begin with detailed pictures of the people who constitute the community, portraits that show how individuals are defined through their relationships with others. Following Riehl, Eliot calls this sort of collective portraiture "the natural history of our social classes . . . [showing] the degree in which they are influenced by local conditions, their maxims and habits, . . . [and] the interactions of the various classes on each other."[16]

In natural history Eliot found a way of studying interconnection that remained grounded in individual material beings. Natural history described creatures with such careful attention that it stirred up sympathy and admiration, inspiring in readers a kind of benevolent desire for all its subjects to flourish. As the literary historian Amy King has documented, the longstanding relationship between natural history and natural theology tinged this mode of scientific writing with religious awe for much of the nineteenth century.[17] Eliot was well-versed in this aspect of natural history writing. She had read the parson-naturalist Gilbert White's classic *Natural History of Selborne* (1789) the year before her Ilfracombe excursion, at around the same time that she was perusing contemporary histories of science by Herbert Spencer and William Whewell. She would remain an avid reader of natural history writing in the years to come.[18] As Eliot saw it, the naturalist's reverent attention to other creatures differed from the cold-blooded analysis of other sciences, and it thereby provided a model for the artist's primary role—to offer "a picture of human life . . . [that] surprises even the trivial and the selfish into that attention to what is apart from themselves, which may be called the raw material of moral sentiment."[19] Where political economy had ossified into a set of assertions and equations attesting to its own scientific authority, natural history remained dedicated to the messy specificity of actual *"life, which is a great deal more than science."*[20]

In fact, a primary attraction of natural history lay in its ability to draw attention to the gap between the certainties of science and the mysteries of being alive. Because natural history named both a kind of writing and a kind of experiential learning through engaging with other creatures, it made palpable the discrepancies between tidily written accounts of community and the practical challenges of interacting with other beings. Eliot and Lewes experienced these discrepancies firsthand on their holiday as they hunted among tide pools, trying to collect, identify, and keep alive the marine invertebrates they brought home to their aquarium. Nor were they alone: the fascinations and frustrations of this process were becoming an increasingly familiar element of middle-class experience. The rage for aquarium-keeping sweeping Britain at the time had led many amateurs to realize how difficult it was to build and maintain a functioning community of even the simplest sea creatures. In the struggle of frustrated hobbyists to replicate the self-sustaining complexity of the tide pool, Eliot found an apt image for the failure of abstract social theories to produce real collective cohesion:

> And just as the most thorough acquaintance with physics, or chemistry, or general physiology will not enable you at once to establish the balance of life in your private vivarium, so that your particular society of zoophytes, molluscs, and echinoderms may feel themselves, as the Germans say, at ease in their skin; so the most complete equipment of theory will not enable a statesman or a political or social reformer to adjust his measures wisely, in the absence of a special acquaintance with the section of society for which he legislates, with the peculiar characteristics of the nation, the province, the class whose well-being he has to consult. In other words, a wise social policy must be based not simply on abstract social science, but on the Natural History of social bodies.[21]

In reality, the distinction Eliot drew between the abstraction of economic thought and the practical, passionate engagement of natural history was too simple. There continued to be substantial traffic between economic and scientific undertakings, including the rise of natural history collecting as a serious business in its own right. Aquariums, Wardian cases (a kind of early plant terrarium), and other Victorian technologies of collection and display led to a booming trade in living specimens, one that depended on industrial networks of exchange and encouraged the commodification of living creatures. Dealers classified such creatures and assigned them market values to maximize profits, sometimes overharvesting local populations to serve far-flung consumer demand.[22] And although she could not have known it, Eliot was lavishing praise on the moral aspects of natural history at the same time that Alfred Russel Wallace was halfway across the world in Borneo, cheerfully gunning down orangutans.[23]

The intimate relationship between natural history and political economy that concerned earlier social-problem novelists was clearly still very much alive. Nevertheless, Eliot recognized in nature study a promising model for the social work that realist fiction might perform. Her experiences on the Devon coast showed her that even the professional collecting of other creatures could create intimacy and sympathy that exceeded the calculus of monetary transactions. After a visit to one specimen collector at Ilfracombe, Eliot wrote that she and Lewes "were pleased with his nice daughter, who seems to take a real interest in the animals quite apart from commercial considerations."[24]

In Eliot's eyes, then, the virtues of natural history were manifold. It maintained the personalized approach to communal relations that economics had abandoned. Its loving attention to plants, animals, and the earth stood as an exemplar of the consideration that authors of "our social novels" should extend to all classes of human beings.[25] And its interest in cross-species relationships provided a prototype for the way a good novelist might look at interactions across classes. In each of these cases, however, natural history was only a model, a metaphor for how novelists might write about human society. Nowhere in her review does Eliot dwell on the necessity or value of writing about the relationships that bind human beings to the creatures around them. As soon as she took up writing novels herself, though, the problem of defining humanity in an entangled, interconnected world once again reared its ugly head—and unlike her predecessors, Eliot could not rely on religious faith to resolve it.

ANIMAL CHARACTERS AND HUMAN RELATIONS

Shortly after *Adam Bede* appeared under the name George Eliot, the debut novelist received an unusual piece of fan mail. It came from Jane Welsh Carlyle, the wife of the influential, visionary, and notoriously cranky essayist Thomas Carlyle—the same man who had given the Condition of England debate its name. As Jane Carlyle's letter describes it, Eliot's novel had an extraordinary effect on her, causing a radical expansion of her social sympathies. Carlyle's account of her feelings closely follows Eliot's theorizations about what novels should do for a reader's sense of community. "In truth," Carlyle exclaimed, "it is a beautiful most *human* book!" Immediately after this sentence, however, Carlyle's letter takes an unexpected turn. Her praise for Eliot's vision of humanity is sidelined by an unexpected distraction: the novel's wonderful dogs. "Every *Dog* in it, not to say every man woman and child in it, is brought home to one's 'business and bosom,' an individual fellow-creature!" she raved. "I found myself in charity with the whole human race when I laid it down—the *canine* race I had *always* appreciated—'not wisely but too well!'"[26]

Carlyle was a voracious, perceptive reader. Her remarks were expressed lightheartedly, but they pinpoint a legitimate moral question burbling just beneath the surface of Eliot's work: the question of what kind of sympathy and ethical recognition animals deserve in the novel's imagination of community. How was it possible to love animals wisely and not *too* well? For Carlyle, the dogs of *Adam Bede* were as fascinating and sympathetic as any human characters, and the book's careful attention to them numbered among its many strengths. For Eliot, however, the sympathetic fascination of her fictional dogs was a matter of serious concern. This concern is, in fact, a recurring theme within the novel. In the world of *Adam Bede*, canine characters prove almost irresistible to the people around them—to the point that the animals threaten to divert and displace the sympathies emerging between human beings. Eliot had learned how to attend to other creatures with care and love from natural history, but now that care and love seemed to endanger the social sympathies she was fighting to promote in her realist fiction. Even as she bestowed deliberate sympathetic attention on the animals of *Adam Bede*, she wrestled with how much space they should take up in novels that were supposed to focus on fostering understanding between human beings.

In at least one sense dogs are given undeniable priority in the novel. The first living being to appear in Eliot's book is Adam's dog Gyp. The story begins with a visual survey of a carpenter's workshop that eventually settles on Gyp, "a rough grey shepherd-dog" resting atop some wood shavings.[27] Gyp keeps a watchful eye on the lead craftsman in the shop, the man who turns out to be Adam himself. It is an amusingly letter-by-letter reversal of biblical order: instead of introducing Adam as the subject of the loving ministrations of God, Eliot introduces her Adam as the subject of the loving ministrations of a dog. But Gyp proves more than just an opening gimmick. Over the course of the story, he becomes a character in his own right and a surprisingly central figure in the Bede household. Several other dogs play important parts in the tale, too. Whenever they become prominent, though, they disrupt the social relations of the novel, and Eliot worries aloud over how much attention they deserve in this complex moral community.

The first hint that canines might snarl the ethical web of Eliot's fiction occurs early in the novel. Adam arrives home from his workshop to discover that his father, who is also a carpenter, has failed to come home to finish a project. Adam correctly guesses that his father stopped by a favorite drinking spot and forgot his duty; now his eldest son will have to work all night to make up for it. The usually even-keeled Adam is furious. When he threatens to leave the family and strike out on his own, his weak-willed mother, Lisbeth, starts sobbing and defending her husband. Through tears she urges Adam to rest and eat the dinner she has prepared. He refuses. Stymied, Lisbeth looks around for another, less judgmental subject for her loving attentions. Lighting on the ever-present Gyp, Lisbeth tries to coddle him in Adam's place, offering him a meal instead.

If *Adam Bede* were exclusively interested in the stories of human beings, the conflict might end there. But Eliot treats Gyp as a real character: he is an agent capable of meaningful action, an individual whose interests and affections are powerful enough to alter the plot.[28] Unlike most human characters, Gyp cannot express his inner self through words. Nevertheless, he manages to participate in all the story's social dynamics. The dog detects that Adam and Lisbeth are fighting, and he even worries that accepting Lisbeth's affection might count as a betrayal of Adam. A standoff ensues. Eliot lingers over the dog as he sits in tense ambivalence, caught in an emotional triangle that now connects Gyp, Adam, and Lisbeth. "Gyp was watching his master with wrinkled brow and ears erect, puzzled at this unusual course of things," Eliot writes in *Adam Bede,* "and though he glanced at Lisbeth when she called him, and moved his fore-paws uneasily, well knowing that she was inviting him to supper, he was in a divided state of mind, and remained seated on his haunches, again fixing his eyes anxiously on his master" (39). The impasse only ends when Adam intercedes—not to make up with his mother but to ease Gyp's suffering. Adam encourages Gyp to run to Lisbeth, and the dog trots off happily for food and fondling.

It may seem ridiculous to spend time rehashing a brief dispute over dog food. After all, the community of the novel is already an imaginary world, and we are accustomed to treating imaginary things—imaginary animals in particular—as beings of little significance. Eliot insists, however, that this fight actually matters. Her attention to the fight is yet another sign that Gyp is more than a prop or a bit of literary scenery. As the critic Alex Woloch has shown, the populous world of the Victorian novel is structured around constant competition between characters for the limited resources of time and space on the page. As characters acquire more of these limited resources, they also acquire fuller, rounder characterization and a greater share of the writer's and reader's sympathetic attention.[29]

When Gyp enters into such competition, then, he proves his own character status—something writers rarely afford to nonhuman animals. More importantly, the competition is realistic, and it has analogues and impacts in the real world. In the contrast between Adam's treatment of his mother and that of his dog, Eliot sees a telling example of a much larger and more problematic pattern of human behavior. "Adam noticed Gyp's mental conflict," she muses in the novel, "and though his anger had made him less tender than usual to his mother, it did not prevent him from caring as much as usual for his dog. We are apt to be kinder to the brutes that love us than to the women that love us. Is it because the brutes are dumb?" (39). As it turns out, the question is purely rhetorical. It leads Eliot into an extended defense of Lisbeth's excellent qualities, qualities that persist despite her aggravating, stereotypical habits of chattering, crying, and complaining. Adam softens toward his mother before long, and their relationship recovers.

Eliot, by contrast, never really recovers from this early interspecies conflict. The competing sympathies introduced by animal characters remain a major concern for her as a novelist. Granted, this concern is eclipsed by the acts of infidelity and infanticide that take center stage for most of the novel. When the time comes to tie up loose ends in the novel's conclusion, however, Eliot feels compelled to revisit the triangular relationship between Adam, Lisbeth, and Gyp; for her, it still represents a meaningful problem that needs to be resolved. Left to themselves on an easygoing Sunday, Lisbeth and Adam relax into their happiest domestic roles. Adam reads quietly in his Bible. Lisbeth plays the part of the doting mother cooking for her son. Gyp figures in this felicitous scene, too—only now he is the outsider, not Lisbeth. Eliot prioritizes Lisbeth by picturing this scene through her often-neglected perspective. Eliot describes how "the smell of roast meat before the clear fire in the clean kitchen, the clock ticking in a peaceful Sunday manner, her darling Adam seated near her in his best clothes, doing nothing very important, so that she could go and stroke her hand across his hair if she liked, and see him look up and smile, while Gyp, rather jealous, poked his muzzle up between them—all these things made poor Lisbeth's earthly paradise" (445). For Lisbeth, a crucial part of the happiness of this ending is that the novel's questions of sympathy and moral priority have been decided in her favor, not in Gyp's. In case there is any remaining doubt, however, Adam says the quiet part out loud. Glancing up from his Bible, he wryly notes Gyp's jealousy and then affirms the dog's subordinate position in the household. "Why, mother," he observes, "thee look'st rare and hearty this morning. Eh, Gyp wants me t' look at him: he can't abide to think I love thee the best" (446). All is well, and the hierarchy of humans over dumb brutes has been restored—in the Bede family, at least.[30]

The dynamic between dogs and people plays out differently in other corners of this fictional world. Adam's character has an eccentric, dark double in the figure of the schoolmaster Bartle Massey. Like Adam, Massey is a single man with an intense attachment to his dog. Unlike Adam, however, Massey has no women at all in his life—excepting his "brown-and-tan-coloured bitch" Vixen (215). As far as Massey is concerned, Vixen may as well be his wife. "He always called Vixen a woman," Eliot notes, "and seemed to have lost all consciousness that he was using a figure of speech" (215). At times Eliot plays along, noting for example that "[t]he table was as clean as if Vixen had been an excellent housewife in a chequered apron" (216). Adam, too, seems to delight in this canine charade. Seeing that Vixen has given birth to a litter of puppies, he exclaims, "Why, you've got a family, I see, Mr Massey?" (215).

Yet the amusing exchanges about Massey's relationship cannot hide its more disturbing features. Vixen's primary role in Massey's life is to serve as the dumping ground for a never-ending torrent of misogynistic abuse. The name he gave her is itself a demeaning term for a woman. As Massey talks to and about Vixen, he strings together slurs and fantasies of violence that make no clear distinction between dogs and female human beings: "If I'd known Vixen was a woman, I'd never have held the boys from drowning her. . . . And now you see what she's brought me to—the sly, hypocritical wench. . . . I've wished again and again I'd been a bloody-minded man, that I could have strangled the mother and the brats with one cord" (215). The charitable Adam can only listen for so long before suggesting that Massey should be kinder to "the creaturs God has made to be companions for us." He intends the phrase to apply to women, but it works just as well for companion animals like Vixen. His mild rebuke launches Massey into pages of invective against women, where he compares them to "adders and wasps, and bugs and wild beasts" (218). His tirade only pauses when Vixen, picking up on his fury, barks sympathetically. Massey immediately redirects his anger from women in general to Vixen in particular: "'Quiet, Vixen!' snarled Bartle, turning round upon her. 'You're like the rest o' the women—always putting in *your* word before you know why'" (218).

Bartle Massey presents a worst-case scenario of animal companionship. His adoption of Vixen saved her life, which seems like a positive moral outcome. But adopting Vixen also saved Massey from the universal moral obligation to cultivate sympathy toward other members of the community—and it is women, here, who pay the price. The playful pretense that Vixen is Massey's common-law wife masks a deeply unsettling reality. Vixen really has taken the place of a human companion who might have lived with Massey, someone who could have taught him to better understand the ways of women and the importance of treating other people justly. As a result, Massey remains a stunted misogynist, safe in the knowledge that his dog will silently cling to him without challenging his outlook on women and the world. Near the end of the book, when Adam's lover Hetty Sorrel faces the gallows for her crimes and Adam has been forced to acknowledge his own moral limitations, Massey remains unreformed. He is unable to see how Hetty's life might matter, and he continues to link her worthlessness to her womanhood: "For my own part, I think the sooner such women are put out o' the world the better.... What good will you do by keeping such vermin alive? eating the victual that 'ud feed rational beings" (375).

Massey's essentially Malthusian question—returning to the issue of who eats what, and at whose expense—recalls the competition over food between Lisbeth, Gyp, and Adam early in the novel. In Eliot's writing, however, the scarce resource of concern is rarely food or money. Such shortages still exist in her work, and they do play a structural role in her plots. Eliot is, after all, part of the tradition of ecological plotting traced in this literary history, and the circulation of matter and money continues to undergird the stories of community she tells. At bottom, for example, *Adam Bede* is the story of a carpenter whose dedication to both the matter of wood and the fabric of his community lead him to consider a career shift to forestry. The plot of Eliot's next novel, *The Mill on the Floss*, turns entirely on the confusing relationship between a community's shared natural resources (water) and its more private resources (land and capital). The plot of *Middlemarch* similarly hinges on monetary networks of debt and inheritance, both of which are tangled up in various characters' dealings in horses, railroads, and agricultural land.[31]

These material and ecological underpinnings are easily forgotten, however, because Eliot herself obscures them. According to her logic, any response to material privation must first tackle the underlying issue of our hoarded resources of individual sympathy. Because solutions to communal problems require a kind of fellow feeling often in short supply, it is the circulation of moral energies, rather than material ones, that becomes Eliot's main preoccupation. The pattern that Malthus theorized and that the social-problem novelists decried—namely, that free-market exchanges lead to animals consuming resources that could feed and clothe needy human beings—was not the kind of competition that bothered Eliot most. It was animals' tendency to consume the moral resource of humane sympathy that made them so problematic in her fiction. When Adam lets his anger obscure his love for his mother but not for his dog, his behavior demonstrates a potential obstruction to the flow of fellow feeling necessary to improving society. Massey's relationship with Vixen presents a more advanced case of the same moral pathology. Where Adam lets his dog siphon off some of the affection needed by another human being, Massey replaces all women in his life with a single long-suffering animal.

None of this meant that animals were necessarily less worthy of sympathy and moral inclusion than the humans around them. For Eliot, the danger of animals lay in their unknowability, the difficulty of ever fully accessing their particular kinds of internal life.[32] Like most nonhuman animals, the dogs of *Adam Bede* possess modes of expression that remain opaque to us. This trait is exaggerated in the case of Gyp, who lacks not only human language to communicate but even the simpler apparatus of a tail. "If Gyp had had a tail he doubtless would have wagged it," Eliot remarks shortly after the dog's introduction, "but being destitute of that vehicle for his emotions, he was like many other worthy personages, destined to appear more phlegmatic than nature had made him" (11). Gyp's curtailed ability to express himself makes him an object of extended metaphysical musing. The saintly Dinah Morris, a woman whose fellow feeling is so profound it amounts to "sympathetic divination," at once pities Gyp and admits she is uncertain whether it makes sense to do so: "'Poor dog!' said Dinah, patting the rough grey coat, 'I've a strange feeling about the dumb things as if they wanted to

speak, and it was a trouble to 'em because they couldn't. I can't help being sorry for the dogs always, though perhaps there's no need. But they may well have more in them than they know how to make us understand, for we can't say half what we feel, with all our words'" (107–8).

It is not the animals' fault that humans fall short of understanding them. As Dinah observes, their incomprehensibility does not mean they lack ethically significant internal lives. But their incomprehensibility does make them convenient targets for the projections of the self-centered humans around them. Vixen proves an ideal companion to Bartle Massey, for example, because she cannot talk back and surprise him into the kind of attention to otherness that encourages moral growth. Instead, Vixen becomes the perfect vehicle for Massey's misogyny; her silence helps affirm his egotistical ideas without requiring him to do the work of interpersonal understanding, the work that might have changed him for the better. In this sense, nonhuman animals provide an extreme example of the moral dangers that can arise in human attachments, too. These same dangers underpin the central relationship of *Adam Bede,* the romance between the thoughtful Adam and his shallow companion Hetty.

As Adam eventually learns, the danger of loving Hetty is virtually identical to the danger of loving a pampered lapdog. In fact, the beautiful but self-involved Hetty is often described in nonhuman terms. "Hetty," Eliot writes, "had the luxurious nature of a round, soft-coated pet animal" (340). She has "a beauty like that of kittens, or very small downy ducks making gentle noises with their soft bills" (76). Her animal allure is enhanced by her silence. Adam's intelligence and wide-ranging interests are alien to someone as simple and egotistical as Hetty. Because Hetty never offers him a glimpse of her actual personality, though, Adam fills her quiet inscrutability with his own imaginings. The loveliness he thinks he discovers in her is really just a projection of the best parts of himself. "How could he imagine narrowness, selfishness, hardness in her?" Eliot asks. "He created the mind he believed in out of his own, which was large, unselfish, tender" (319). Through her very inscrutability, Hetty becomes a perilous site of projection for everyone around her. In this sense, at least, her relationship with Adam is analogous to Vixen's relationship with Bartle Massey. Massey uses Vixen to externalize his hateful image of women, while Adam uses Hetty to externalize his own generosity and kindness. Regardless of whether the false image they create is lovable or despicable, the act leads both men to seriously misread another living creature.

In the end, though, it is not the error of misreading others that most concerns Eliot. After all, Hetty stands to benefit from Adam's misrepresentation, which paints her as a far more admirable person than she is. The problem with animals like Vixen and people like Hetty is that they offer frictionless enticements to connection—relationships that look and feel like real, meaningful encounters with others but that actually deliver only the doubled self-satisfaction of engaging with our own egos. In yet another comparison of Hetty to nonhuman nature, Eliot warns, "We look at the one little woman's face we love, as we look at the face of our mother earth, and see all sorts of answers to our own yearnings" (188).

For Eliot, the entire social and ethical project of the novel involves introducing readers to forms of otherness they never took seriously before. Trying to understand these alien others stirs "the awakening of social sympathies" that makes practical action toward a better world possible.[33] But this introduction of otherness must be gradual. When the distance between ourselves and an other becomes too great, that other becomes unfathomable. As Eliot admits in *Adam Bede*, "[T]o shift one's point of view beyond certain limits is impossible to the most liberal and expansive mind; we are none of us aware of the impression we produce on Brazilian monkeys of feeble understanding—it is possible they see hardly anything in us" (185).

The great danger of ushering animals and other nonhuman beings into our imagination of community lies in their inscrutability, their inability to confirm or deny our understanding of them. It may be—especially in the case of Gyp and other companion animals—that nonhuman creatures are every bit as worthy of ethical consideration as human beings. But the interspecies relationships of *Adam Bede* highlight the fact that many people use nonhumans as convenient confirmations of what they want to see and believe about themselves and about others. Rather than eroding their own selfishness through real, sympathetic relationships with others, these seemingly humane animal lovers expand their own egos by locating cherished parts of themselves in animals who lack any capacity to contradict them. Such encounters insulate people from the shock of engaging with actual otherness. They short-circuit the project of expanding sympathetic connections, the project that powers political and ethical progress.

Eliot's venture into fiction began as a result of her own enchanting encounters with nonhuman otherness. In the course of writing her first novel, however, she started to see the necessity of moving away from it. If the ethical project of the novel was to succeed, she would have to find some means of maintaining its emphasis on interconnection while also driving nature and society apart. The social-problem novelists had faced a similar issue. They found a way to reassert human priority by embracing the divinity of the human soul as essentially more important and more real than the material ecologies of interconnection. The resolutely secular Eliot did not share those novelists' faith in the immortal human soul.[34] Yet she too would turn away from materialism in her attempts to partition humanity from nature, relying not on spirit but on something almost equally ethereal: literary form.

GROWING APART FROM NATURE

Eliot's second novel, *The Mill on the Floss*, is a book-length exploration of how literary form can drive humanity and the natural world apart. On the face of it, the book is primarily a coming-of-age story. It centers on Maggie Tulliver, a spirited girl whose defiance of social conventions often lands her in hot water. As Maggie matures and enters adult society, the pressure to conform to the expectations of her rural English town becomes more and more stifling until she is decisively expelled from her community. Most readers and critics interpret the clashes between Maggie and those around her in terms of sex and gender. The novel plays up the fact that Maggie's passion and intelligence defy the conventions of the dim, demure, marriageable woman desirable at the time.[35] Her unconventionality is central to her appeal as a heroine, and it suggests the damning limitations of her conservative society. Yet Maggie also causes social unease because she is wild in a more fundamental sense. From the very beginning, Maggie shows signs of being too animal to belong in human social groups.[36] She maintains perverse affiliations with nature throughout the novel, even as those around her grow increasingly invested in more narrowly human forms of community. Beneath the novel's explicit condemnation of sexual double standards, then, there is a more basic but also more ambivalent message that full maturity requires growing away from nature. That message is driven home through the counterexample of Maggie, the stubbornly wild child who refuses to listen to it—and ends up paying a heavy price for her mistake.

At the beginning of the novel, Maggie is beastly in the most delightful ways. She is "a wild thing" in every sense, a "small mistake of nature" whose appearance and behaviors consistently call the nonhuman to mind.[37] When she tosses "her mane" to keep her hair out of her eyes, Maggie has "the air of a small Shetland pony" (13). When she prepares to defend her beloved brother, Tom, Maggie takes on the watchful look of "a Skye terrier suspecting mischief" (15) and, later, "the eyes of a young lioness" (200). Maggie's inhuman wildness draws her into conflict with the adult world almost instantly. She introduces inappropriate subjects when talking with visitors. She cuts her own hair to disastrous effect. She pushes her sweet little cousin Lucy into the mud, scandalizing the relatives who have "so often told [Maggie] she was like a gypsy, and 'half wild,'" that she decides to fulfill her destiny by running away from home (99). Although the word "gypsy" is no longer considered an accurate or appropriate name for Romani peoples, its usage here is significant. It drives home both Maggie's status as an outsider to English social customs and the way her marginal position aligns her with Gyp the dog, whose name calls up the same stereotype of the gypsy as an outcast who proves simultaneously threatening and alluring.[38]

The Mill on the Floss opens with an almost parental assurance that Maggie will not remain an alluring outcast forever. If all goes according to plan, she will grow out of her wild behavior and animal attachments as she approaches adulthood. Eliot makes allowance for Maggie's untamed childhood by ascribing similar affinities for the natural world to all the children in the novel. Social proprieties do not yet apply to them, and they are "used to hear[ing] themselves talked of as freely as if they were birds, and could understand nothing, however they might stretch their necks and listen." When Maggie, her brother, and her cousin are liberated from adult supervision, they run off "with the alacrity of small animals getting from under a burning-glass" (65). Everyone in the town of St. Ogg's appears to share the opinion of Lucy's father, Mr. Deane, who maintains that "what goes on among the young people is as extraneous to the real business of life as what goes on among the birds and butterflies" (397).

Whatever their individual differences, then, Maggie, her brother, and their cousin Lucy all begin life in a wild, pre-socialized state. Eliot spells out this fact in a lengthy aside when Tom and Maggie fight. Here, Eliot foreshadows how time will humanize these little creatures and, in the process, alter their relationship to each other and to the wider world: "We learn to restrain ourselves as we get older. We keep apart when we have quarrelled, express ourselves in well-bred phrases, and in this way preserve a dignified alienation, showing much firmness on one side, and swallowing much grief on the other. We no longer approximate in our behaviour to the mere impulsiveness of the lower animals, but conduct ourselves in every respect like members of a highly civilized society" (37).

The Mill on the Floss endorses this time-honored process of humanization—sort of. It is, after all, a bildungsroman, the German term for a novel of education or coming-of-age story. As the plot unfolds, Maggie grows up both literally and figuratively, learning difficult lessons about money, gender, sex, and disability along the way. That much is typical: establishing connections between the literal act of growing up and a more figurative form of maturity is central to the cultural work of the bildungsroman as a genre. The standard bildungsroman is not just a matter-of-fact narrative about events in an individual's life as the years pass. It is also didactic. Like a schoolteacher, it imparts lessons that instill in readers what socially acceptable behavior looks like. While most coming-of-age stories in the English tradition have happy endings, the rewards of maturity appear only after the main character has let go of certain dreams, values, or beliefs associated with the idealism of youth. The coming-of-age story is overwhelmingly about the importance of taming or mastering one's own desires in order to join society and reap the benefits of belonging. For that reason the English bildungsroman is widely understood as a genre that focuses not just on rites of passage but also on the art of compromising with existing values, a genre that plays up the many benefits of such conformity.[39]

Eliot herself never seems totally at ease with such compromise, however. The sections devoted to Tom and Maggie's unsocialized childhood are easily the happiest portions of the novel. In the passage above, it is actually the inhumanity the siblings share that ensures they will make up and end their fight happily: "Maggie and Tom were still very much like young animals, and so she could rub her cheek against his, and kiss his ear in a random, sobbing way; and there were tender fibres in the lad that had been used to answer to Maggie's fondling; so that . . . he actually began to kiss her in return" (37). Sadly, their overt love must go underground if Tom and Maggie are to internalize the "dignified alienation" that separates human beings from not only the "impulsiveness of the lower animals" but also from each other. Eliot shows a certain skepticism, too, about the depth of such transformations. People do not so much *become* civilized as learn to act "*like* members of a highly civilized society."

But this skepticism is ambivalent; it cuts both ways. If Eliot is unconvinced that the benefits of socialization outweigh the costs, she has reservations about the animalized, pre-social state of being, too. "[T]he mere impulsiveness of lower animals" is not exactly a desirable condition. Eliot even slips into an uncharacteristic expression of shame as she narrates the end of Tom and Maggie's conflict: "[T]hey ate together and rubbed each other's cheeks and brows and noses together, while they ate, with a humiliating resemblance to two friendly ponies" (37). Who exactly is humiliated here? Not Tom and Maggie, whose wild state leaves them totally unselfconscious as they revel in the moment. The only person who feels any shame over this exchange is Eliot herself, who has been socialized enough to see how its animalistic affection violates the norms of grown-up human interaction.

The entire novel is riddled with this ambivalence. On the one hand, Eliot feels the nagging importance of inserting some distance between nature and society if she is going to properly channel her readers' sympathetic energies toward communal moral improvement. On the other hand, she knows that any distance she succeeds in creating will be, at a philosophical level, artificial and even arrogant. The act of prizing human society apart from the larger natural communities we belong to is finally just "dignified alienation," little more than a pose we assume to turn our backs on the fundamental reality of interconnection.

In *The Mill on the Floss*, Eliot leans into the generic structure of the bildungsroman to nudge Maggie along, encouraging her (and us) to hurry up and grow away from the nature that fascinates us as children. The novel's attempt to enforce divisions between nonhuman nature and human culture is, in fact, a widespread but rarely remarked feature of the bildungsroman as a literary institution. In a compelling argument, the critic Helena Feder has observed that modern coming-of-age stories are overwhelmingly about "the formation of the human itself. The bildungsroman is humanism's story of becoming human as becoming part of culture, the humanist origin story of culture itself, of its self-creation out of nature."[40] Maggie is an exceptionally divisive and problematic figure in this tradition. The story's relentless march through time requires her to put aside childish things—including her affiliations with nature—in order to accede to the adult world of civilized human beings. "But that is the trial I have to bear in everything," Maggie herself laments at one point in the novel. "I may not keep anything I used to love when I was little. . . . It is like death. I must part with everything I cared for when I was a child" (279).

Yet in an unexpected twist, Maggie proves too much for Eliot herself. Headstrong to the end, she refuses to accept what both the plot and Eliot insist she must do. Maggie retains her kinship with the nonhuman even in those periods when she shows the most mastery over her younger, wilder self. Her enduring wildness is a source of fascination for her fellow outsider Philip Wakem, a classmate of Tom's who meets Maggie just before she leaves for boarding school. There he finds himself enthralled by "Maggie's dark eyes[, which] remind him of the stories about princesses being turned into animals" (167). Later, when compounding circumstances cause the Tullivers to lose their respectable position in society, Maggie manages to reenter the social world only as a sort of exotic companion animal to her rich cousin Lucy. Lucy, Eliot tells us, "was fond of feeding dependent creatures, and knew the private tastes of all the animals about the house" (342). When Maggie is taken under Lucy's roof, she is overawed at her cousin's hospitality. Seeing Maggie again under these changed circumstances, Philip immediately identifies the beastly logic of Lucy's generosity. "You must be better than a whole menagerie of pets to her," he says to Maggie (380). At heart, however, Maggie is still a wild thing unfit for the drawing room. It won't be long before she bites the hand that feeds her.

When Maggie accidentally betrays the trust of her cousin, her social fate is sealed. Public opinion in the town of St. Ogg's turns decisively against Maggie. Expelled from the family home, she takes lodgings with the oddball Bob Jakin, another animal lover whose dim understanding of human social doings leaves him confused about why people are upset with Maggie in the first place. But Maggie knows that she has effectively cut all connections to the social network around her: "[S]he had rent the ties that had given meaning to duty, and had made herself an outlawed soul" (436). She possesses "unwomanly boldness and unbridled passion" (455), the first trait marking a violation of gender conventions, the second a refusal to tame her inner animal. Her behavior sets a terrible example for other up-and-coming members of society, so it is finally for "the preservation of Society" (454) that Maggie must be excluded from it, "else what would become of Society?" (455).

Maggie's failure to comply with standard proprieties is a problem that has implications for both social conventions and literary ones. At a social level, Maggie defies those conventions that ask her to speak and behave in the ways recognized as admirably feminine and human. Her failure to do so deprives her of the possibility to lay claim to those traits, rendering her unwomanly, beastly, and wild. At a formal level, Maggie defies those conventions of the bildungsroman that ask her to tame herself as a character to achieve some semblance of a happy ending. Her failure to do so deprives her of the possibility to claim such an ending—and even of the possibility to continue on as a character in the story. "I will sanction no such character as yours," Tom tells her in a particularly sanctimonious moment following her public shame (449).

Here as elsewhere in the novel, Maggie's brother Tom turns out to be at once excessively strict and unnervingly correct. By the end of the story, Maggie's fate will be to lose her standing as a character entirely. In what the critic Nathan K. Hensley calls "the most famous surprise ending in Victorian fiction," Maggie drowns, caught up in one of the periodic floods that strike the region—a conclusion that has long infuriated the novel's readers.[41] Shocking as it is to find a coming-of-age story that ends with the death of its beloved heroine, there is something sweet and fitting about Maggie's death, too. As Eliot herself notes, the death effectively returns Maggie to the natural scenery of her childhood, a place that includes all the beloved associations the bildungsroman asks her to give up. When Maggie sinks to her death, she drags her brother down with her. In their last conscious thoughts they end up "living through again in one supreme moment the days when they clasped their little hands in love, and roamed the daisied fields together" (483). The epitaph on the siblings' tomb (484) is itself a return to the beginning; it is identical to the novel's opening biblical epigraph, "In their death they were not divided." Within the story, the epitaph clearly refers to the estranged Tom and Maggie, united finally by the tragic fate they share. But once again, the explicit social content of the novel is tightly interlocked with the subtler mechanics of its literary form.

The epigraph and epitaph that bookend *The Mill on the Floss* apply not only to brother and sister but also to character and setting. In death, Maggie is no longer divided from her beloved earth: she returns in the most literal sense to the natural world she could never bring herself to renounce. Extinguished as a character, she merges with the novel's scenic background. Looking more closely at her abrupt migrations across the invisible line that separates character from setting usefully clarifies the rarely discussed formal means of differentiating the two. It also demonstrates how Eliot and other novelists came to use the stylistic power of language to distinguish character from setting, calling on literary form to shore up artificial barriers between human society and the nonhuman world that is so hopelessly intermingled with it.

SETTING NATURE ASIDE

The landscapes of *The Mill on the Floss* are so boring they put even the narrator to sleep. The novel begins with the kind of extended scene-setting many readers expect from nineteenth-century fiction. After several pages cataloguing the surroundings of Dorlcote Mill, however, something unprecedented happens: the narrator nods off. "It is time, too," Eliot writes, "for me to leave off resting my arms on the cold stone of this bridge. . . . ." (8; original ellipsis). Only after the extended ellipsis and a paragraph break does the narrator realize what has happened, explaining the situation to the reader: "Ah, my arms are really benumbed. I have been pressing my elbows on the arms of my chair, and dreaming that I was standing on the bridge in front of Dorlcote Mill, as it looked one February afternoon many years ago. Before I dozed off, I was going to tell you what Mr and Mrs Tulliver were talking about, as they sat by the bright fire in the left-hand parlour, on that very afternoon I have been dreaming of" (8–9).

It is a bizarre opening—at once utterly impossible and strangely sly. An oral storyteller might fall asleep in the midst of talking, but this five-hundred-page novel is not an oral story, and it's never treated as one anywhere else in the book. A writer might nod off in the midst of writing, too, but their sleep would not produce a set of five periods on the page. And no writer would mark their return to wakefulness with a new paragraph opening with "Ah." In its obvious, self-conscious artifice, the dozing episode disrupts what Samuel Taylor Coleridge famously called the reader's "willing suspension of disbelief."[42] The interruption draws attention away from the fictional world being described, focusing it on the reader's experience of encountering written language instead. And in an unforgivable writerly sin, the whole thing highlights just how dull that language actually is. The word-painting Eliot employs is calming but utterly lifeless. It resists attention so effectively that even the narrator can't stay invested in it.

This, then, is a scene of carefully cultivated boredom. Eliot shows that she is fully aware how to make writing uninteresting. She also signals that *The Mill on the Floss* will experiment with style and form as tools to gain deliberate control over readerly attention, channeling it away from places where it is less necessary and diverting it toward those for whom sympathy and understanding are in short supply. Like the water whose flows, stoppages, redirections, and floods shape the plot of the novel, sympathy proves to be a vital but dangerously unpredictable resource. It must be carefully managed if it is to serve the needs of the entire community.

The first chapter of the novel displays how to drain a scene of sympathetic attention entirely, distancing it from the main concerns of the narrative. It offers an extended exercise in the formal creation of background, a static setting that throws the human plots of the novel into sharp, intriguing relief. Maggie Tulliver's movement from this scenic background into the foreground (and back again) reveals how literary methods of distinguishing between foregrounded characters and backgrounded settings actually operate, both formally and ideologically. In Eliot's novels and elsewhere, these distinctions work to uphold otherwise dubious dividing lines between society and nature, between humanity and the nonhuman. But Eliot's work also reveals her own awareness that such distinctions were finally aesthetic and arbitrary—and that different ways of narrating and describing might produce different understandings of ethics that could promote a more ecological worldview.

The most striking feature of the scenic description that opens *The Mill on the Floss* is that it is utterly devoid of story. As the novelist E. M. Forster famously explained, a story is more than a series of events laid out chronologically. A story links those events together meaningfully, connecting them to clarify how one occurrence causes or contributes to another.[43] Assigning cause-and-effect relationships lends significance to an otherwise chaotic stream of experience. It "establishes proportions," as the great theorist Georg Lukács puts it, contributing "order and hierarchy among objects and events" and ending "in a proper distribution of emphasis and in a just accentuation of what is essential."[44] The cause-and-effect relations of storytelling impose meaning, relevance, and interest on events that might otherwise seem insignificant, chancy, and random. For Lukács (and, in fact, for Eliot), description is the exact opposite of such ordered, meaningful storytelling. Lacking causal connections between the things it represents, "description merely levels."[45] In its long lists of loosely connected details, "both the important and the unimportant are described with equal attention," Lukács argues, until "all basic aspects of life are smothered under a blanket of delicately delineated minutiae."[46]

Eliot's scenery in *The Mill on the Floss* does more than just exemplify this aspect of description. She expands on and even relishes it. Very few actions or events occur anywhere in her descriptions. Those that do crop up seem not singular and decisive but stable and recurring. Eliot favors the use of passive voice, linking verbs, and present participles (forms ending in *-ing*) that obscure actors and dilate actions over time, draining them of dynamism to transform the verbs into something like gerunds—verbs that act as nouns to name states of being. The novel's opening description kicks off with a fragment, an ungrammatical sentence totally devoid of independent activity: "A wide plain, where the broadening Floss hurries on between its green banks to the sea, and the loving tide, rushing to meet it, checks its passage with an impetuous embrace" (7). As the description sprawls outward, it slips into parataxis, the rhetorical name for a string of details that assume a list-like structure without ever cohering into any neat sequence of events or set of relationships: "There is a remnant still of last year's golden clusters of beehive ricks rising at intervals beyond the hedgerows; and everywhere the hedgerows are studded with trees: the distant ships seem to be lifting their masts and stretching their red-brown sails close among the branches of the spreading ash" (7). The visual compression Eliot mentions—where distant masts and nearby ash appear part of one undifferentiated visual plane—highlights the lack of space or orderly relation in this flattened, equalizing kind of scenic description. It is difficult to focus long on anything here because the objects of attention blur so seamlessly together. Eliot must interject several times to enjoin us to "Look" or "See" or otherwise maintain our flagging interest in this undifferentiated catalogue of objects and creatures. Finally even the narrator cannot help but doze off, unable to stay engaged in the scene before us (8).

These tactics may be boring, but they do exactly what they are supposed to do. Together they have the effect of distancing the opening chapter and everything it describes from the actual action of the novel. All the plants, animals, and objects included in the scenic description become integrated into a single set piece, an acknowledged but unimportant background clearly detachable from the characters and plot that maintain the reader's interest.[47] For the story to begin, in fact, the narrator must break free from this scenery and start again, rapidly shifting perspective from the landscape around Dorlcote Mill to the inside of the Tulliver home—a new position in which the landscape barely appears, demoted to the status of a background visible only through the framing of the window.

Tellingly, Maggie starts out as part and parcel of this distant background. Her position there prompts the first of Mrs. Tulliver's repeated complaints about her daughter's intimacy with the river: "I don't know where she is now. . . . Ah, I thought so—wanderin' up and down by the water, like a wild thing: she'll tumble in some day" (12). For Maggie to even become a character, she must be knocked out of this scenic background and restored to the foreground. And that is exactly what her mother does, knocking hard on the pane of glass to urge her daughter away from the setting and into the story: "Mrs Tulliver rapped the window sharply, beckoned, and shook her head,—a process which she repeated more than once before she returned to her chair" (12).

The scenic description of Eliot's opening provides a textbook example of what have become the conventional techniques for representing nature. It exhibits all the key features of what the literary theorist and philosopher Timothy Morton calls "ecomimesis," a set of strategies typical of nature writing. Together, the techniques of ecomimesis work to convince the reader of the writer's close attunement to the nonhuman environment. Ecomimesis, as Morton explains, treats the narrator as embodied in a particular place, "bringing us into a shared, virtual present time of reading and narrating." It also typically deploys "the paratactic list; the imagery of disjointed phenomena surrounding the narrator; [and] the quietness (not silence, not full sound) . . . that evoke[s] the distance between the hearer and the sound source," which all work together to conjure up "a vivid evocation of atmosphere."[48] The entirety of Eliot's opening occurs within the rhetoric of ecomimesis, a natural scene overlaid with "a great curtain of sound" that includes "[t]he rush of the water, and the booming of the mill" (8).

Yet from the start *The Mill on the Floss* provides reminders that this is not the only way of representing nature. It is an arbitrary artistic choice, one with its own set of emotional and ethical consequences. By having her own narrator nod off, Eliot acknowledges how this kind of writing flattens whatever it depicts, coating it all with the "monotony and tedium" of description (to return to Lukács's words), removing it from "the mesh of human destinies" where things, events, and people "interpenetrate or reciprocally effect each other."[49] Elsewhere in the novel, Eliot uses examples from visual art to contrast this way of depicting the natural world as a static background with a more engaging art of individuated nonhuman figures. When Tom meets the still-young Philip Wakem at their boarding school, for example, Tom is entranced by Philip's realistic sketches of "a donkey with panniers—and a spaniel, and partridges in the corn." These amusing representations of animals in their varied relations to human beings naturally fill Tom with excitement. "O my buttons!" he exclaims, "I wish I could draw like that. I'm to learn drawing this half—I wonder if I shall learn to make dogs and donkeys!" (152). But instead of focusing on the fascination of individual animals, Tom

discovers, "to his disgust, that his new drawing-master gave him no dogs or donkeys to draw, but brooks and rustic bridges and ruins, all with a general softness of black-lead surface, indicating that nature, if anything, was rather satiny." By the end of his training, Tom knows only how "to represent landscape with a 'broad generality,' which . . . he thought extremely dull" (157).

The resemblance between these vague, soporific landscapes and the opening of *The Mill on the Floss* is obvious. Why Eliot would so openly criticize such techniques and then employ them anyway is a harder question. Part of the answer lies in changing understandings of the individual's relationship to the environment in the Victorian era. As the critic Jayne Hildebrand has shown, Eliot was a central figure in the circle of British intellectuals who helped popularize the notion that an individual's natural and social context was best understood not as a loose group of objects, forces, and circumstances but as a unified, monolithic medium—a single, homogeneous environment. This way of understanding the environment came from the French sociology of Auguste Comte, and it constituted a very different way of conceiving interspecies relationships than the methods being developed contemporaneously by Darwin.[50] But Eliot's decision to deploy this new notion of a consolidated environmental surround despite its boring, homogenizing effects has an additional, complicating explanation more localized to the act of novel writing. It was grounded in her ongoing concern with managing the limited reserves of readerly sympathies to ensure they were channeled in the most effective, most necessary directions.

Paradoxical as it may sound, boring backgrounds could play a valuable role in the novel's moral mission. Eliot had been thinking through the importance of distinguishing foreground from background since even before she began writing novels herself. She had reviewed the third volume of John Ruskin's *Modern Painters* (1856) shortly before her trip to the Devon coast. There, she dwelled at length on the art critic's concerns over the proper use of backgrounds in visual art. Ruskin was particularly upset by the painter Paolo Veronese's interpretation of the *Supper at Emmaus* (ca. 1560), where the portrayal of Christ's return from the grave is depicted as "a background to the portraits of two children playing with a dog."[51] For Ruskin, this aesthetic decision was a sign that Veronese had let his painterly ability to capture all kinds of subjects get in the way of his moral responsibility to draw attention to what was most important. "*All* the truths of nature cannot be given," Eliot wrote in her paraphrase of Ruskin's point; "hence a choice must be made of some facts which can be represented from amongst others which must be passed by in silence."⁵² Eliot would reenact Ruskin's argument in the plotting of *The Mill on the Floss,* which opens with the adult world as a background to portraits of two children playing with a dog. Over the course of hundreds of pages the novel rights itself, shifting its perspective to provide a picture of adult society that decisively relegates children and their animal companions to the background.

Maggie Tulliver is the unwitting victim of this formal shift. Despite her mother's repeated attempts to pry Maggie loose from the setting and call her into the human community, Maggie only grows more deeply rooted in the landscape as she ages. As a wild young girl her favorite relative is her aunt Gritty Moss, a farm woman whose name is about as earthy as they come. Even Maggie's mature moments of renunciation only expand her terrestrial associations until she becomes almost planetary. This metamorphosis first appears in book 3 of the novel, "The Valley of Humiliation." The book's name explicitly references the embarrassment of the Tulliver family, as accruing debts and legal fees plunge them into bankruptcy. But "humbling" and "humiliation" both come from the Latin *humus,* meaning ground, earth, or soil; to be humiliated is literally to be forced back to the earth. Fittingly, these chapters show Maggie starting to resemble not just animals but the landscape they populate. Her early attempts at self-discipline turn her into an oddly geological being, a person whose newly stony demeanor hides "fits even of anger and hatred . . . [that] would flow out over her affections and conscience like a lava stream" (266).

These "volcanic upheavings of imprisoned passions" provide hints of the many ways in which Maggie will eventually merge with the earth (272). Before long her positive traits take on the same environmental associations. In moments of candor, for example, she proves "as open and transparent as a rock pool" (411). Over time she begins to resemble not just the ground but the plants that spring from it, too: "With her dark colouring and jet crown surmounting her tall figure, she seems to have a sort of kinship with the grand Scotch firs, at which she is looking up as if she loved them well" (277). As Maggie is "thrown into the background," the distance separating her from other characters becomes at once ideological and aesthetic (327).⁵³ Well-adjusted characters turn toward society and social advancement as they mature, but Maggie perversely faces the other way. A characteristic drawing-room scene shows Maggie avoiding awkward social conversation by "rising and going to the window as if she wanted to see more of the landscape"

(346). It is a distinct yearning toward her surroundings, one so pronounced that Hildebrand has dubbed it a kind of "environmental desire."[54]

Increasingly divided between character and setting, Maggie herself starts to break apart. The formal cracks in her character are most evident in the pictorial space of Philip Wakem's studio. There, Wakem includes pictures of Maggie's face staring out strangely amidst the rural scenes he usually paints, leading visitors to be "startled by a sudden transition from landscape to portrait" (391). A painter of scenery and an outsider himself, Philip sees Maggie's odd position more clearly than anyone else—and he is determined to capture it on canvas. "I've begun my picture of you among the Scotch firs, Maggie," he tells her. "It is an oil-painting. You will look like a tall Hamadryad, dark and strong and noble, just issued from one of the fir-trees, when the stems are casting their afternoon shadows on the grass" (303). The hamadryad, a mythical forest nymph whose life was inextricable from the tree she inhabited, is a fitting symbol for the increasingly earthy Maggie. But whereas Philip imagines Maggie venturing forth from the background, her actual movement is in the opposite direction.

Sundered from the human community and unable to break from her native landscape, Maggie's story ends when the landscape itself comes to claim her. She has been urged, repeatedly, that she could save her social character if she would only "take a situation at a distance," but Maggie objects, saying that she could never become "a lonely wanderer, cut off from the past" (459). One dark night as the rain falls and the water rises, the river that has so long served as textual background suddenly stirs to life. Maggie and her boat are swept out onto the floodwaters, a placeless pre-dawn world where the reconfiguration of land by water is profoundly disorienting. For a brief period of time, the flood scuttles all boundaries between figure and ground, character and setting, animals and landscape as they are rewoven together in a single "curtain of gloom" (479). When the world starts to resolve again into "the slowly defining blackness of objects above the glassy dark," Maggie struggles to get out of the current and return to her native landscape at the mill—a literal rendition of the journey she has spent the entire novel trying to complete in a more figurative way (479). She does finally make it home to Tom, of course. But it is a devastating reconciliation that destroys them both.

As in all tragedies, the destruction of the protagonist is both a painful release and a healing one. Maggie's end restores a certain necessary order.

Always deeply animal, Maggie finally assumes the role of sacrificial lamb. Innocent and overburdened, she is expunged from the novel. Her death makes it possible for Eliot to exorcise the sympathetic pull of the nonhuman world once and for all. *The Mill on the Floss* is Eliot's monument to the difficulty of setting aside our ties to that world, the pain that comes when we are forced to focus on more narrowly human social issues. But it is also a testament to Eliot's determination to go ahead with that breakage anyway, transforming the novel into the secular, humane, and exclusively human art form she thought the world needed most.

HUMANIZING THE NOVEL

The Mill on the Floss marked the end of Eliot's serious engagement with the nonhuman world.[55] By the time she published her final novel, *Daniel Deronda* (1876), Eliot was investing so little time in animals that she accidentally changed the sex of one character's dog over several chapters.[56] But even that level of attention to the nonhuman grew to be unusual in her work. The nonhuman surround of her later novels appears almost exclusively in the form it takes at the beginning and end of *The Mill on the Floss*: as part of the scenic background, the setting that throws human doings into sharp relief. And why not? As Eliot discovered, focusing on human doings made moral stories much easier to tell. Flattening nature into place offered a useful way of curtailing the overwhelming extensiveness of the interspecies community, the sublime sprawl of interconnection that the ecological plot had made thinkable. Scenic description helped authors consign certain beings to the background. It produced a more manageable, narratable community where sympathetic attention could be targeted to address the most pressing ethical needs. As Eliot would write in *Middlemarch*—sounding, again, like Ruskin—a novelist must neglect most of the endless network of related fates to produce a comprehensible picture of society: "I at least have so much to do in unraveling certain human lots, and seeing how they were woven and interwoven, that all the light I can command must be concentrated on this particular web, and not dispersed over that tempting range of relevancies called the universe."[57]

Expelled from the real-life interactions of Eliot's social novels, animals linger around the textual margins, where they exist only as metaphors that offer analogies and images for more central human concerns. By the end of her career these metaphorical animals come to stand in for the necessity of exclusion itself. They become symbols of the impossibility of touching on all relevant human concerns, much less the needs and activities of the many denizens of the nonhuman world. In perhaps the most famous passage from *Middlemarch,* Eliot draws on plants and animals to justify those exclusions necessary to effectively navigate the world. "If we had a keen vision and feeling of all ordinary human life," she writes, "it would be like hearing the grass grow and the squirrel's heart beat, and we should die of that roar which lies on the other side of silence."[58] In this unforgettable simile, the exclusion of nonhuman life is doubled. Not only are the nonhumans here purely metaphorical, but they are brought in to symbolize what obviously *must* be excluded to train our energies and attentions to their proper ends. The metaphor, in short, emphasizes how utterly necessary it is to ignore animals and plants—and uses that necessity to justify further exclusions of certain human beings from our still untutored, slowly growing sympathies.[59]

This perceived tension between comprehensive expansion and moral concentration was not confined to George Eliot. Take, for instance, the case of Henry James, one of Eliot's many admirers. For James the formal question of how to create a sense of coherence and closure in a never-ending network of connections exerted an almost mathematical fascination. Writing a novel, he explained in a preface composed in 1907, is an extended artistic struggle to impose the fiction of a bound and centered circle on what is actually an unbroken web of relationships: "Really, universally, relations stop nowhere, and the exquisite problem of the artist is eternally but to draw, by a geometry of his own, the circle within which they shall happily *appear* to do so. He is in the perpetual predicament that the continuity of things is the whole matter, for him, of comedy and tragedy; that this continuity is never, by the space of an instant or an inch, broken, and that, to do anything at all, he has at once intensely to consult and intensely to ignore it."[60]

Of course, not every Victorian novelist could be as articulate and agonizingly self-aware as Eliot or James. It's hard to imagine Anthony Trollope worrying over these issues, for example. He was too busy filling his quota of three thousand words per day in the early mornings before he left for his job at the postal service. Trollope was one of the countless hardworking authors collectively responsible for the constant stream of entertaining stories that circulated among Victorian readers. These were the authors who produced the vast bulk of the era's fiction. From the perspective of such workmanlike writers, novelists would be ill served wasting time wringing their hands over highfalutin philosophies and artistic choices. "Let their work be to them as is his common work to the common laborer," Trollope advised. "No gigantic efforts will then be necessary."[61] For most novelists, in short, the artistic inclination to privilege the human and to include plants and animals only as scenic backgrounds probably felt natural—a matter of course. Their habits followed more or less directly from the spiritual elevation of the human that helped tie up the loose ends of so many social-problem novels. The primacy of the human reflected a long-standing and commonplace moral consensus, and it offered a path of least resistance to the average writer of realist fiction.

What makes Eliot so instructive is that she ended up ratifying this moral consensus for the novel, but only after a period of intense experimentation and doubt. That doubt led her to much more self-conscious and much more instructive uses of the fundamental categories of character and setting than the average novelist employed. It also left Eliot with a lasting awareness that the divisions she came to rely on were formal and finally arbitrary. Eliot knew that her association of character with humanity did not reflect some transcendent truth about humans and their centrality in the wider world. In *The Mill on the Floss,* for example, the boredom of the opening scenes that put the narrator to sleep is undercut by the catastrophically active scenery at the end of the novel. This internal contradiction emphasizes the danger of settling too comfortably into the idea of a passive natural world—the very idea encouraged by the art of scenic description.

And so, to the very end, Eliot remained conscious of the artificiality of her own exclusions. She wrote as if she were holding open a door for another, better future, a time when hastily excluded others might be readmitted again to novelistic sympathies. This outlook persists even in *Daniel Deronda*. At one point in that story, Deronda joins a party touring the Abbey, a converted medieval monastery that now serves as the estate of Sir Hugo Mallinger. Sir Hugo adopted and raised Daniel when he was a baby, so Deronda knows the landscape well. As they explore the Abbey, the group stops in the estate's stables. Ironically, the housing for horses proves to be "the finest bit of all. . . . It is part of the old church."[62] Eliot launches into a long, dense paragraph of scenic description that once again flattens animal lives, architectural detail, and landscape into a single compressed plane of prose:

> Each finely-arched chapel was turned into a stall, where in the dusty glazing of the windows there still gleamed patches of crimson, orange, blue, and palest violet . . . a soft light fell from the upper windows on sleek brown or gray flanks and haunches; on mild equine faces looking out with active nostrils over the varnished brown boarding; on the hay hanging from racks where the saints once looked down from the altar-pieces, and on the pale golden straw scattered or in heaps; on a little white-and-liver-colored spaniel making his bed on the back of an elderly hackney, and on four ancient angels . . .[63]

The remnants of divinity in this space stir some instinctive respect in Deronda, and he removes his hat automatically. Seeing this, his selfish rival Henleigh Grandcourt mocks him:

> "Do you take off your hat to horses?" said Grandcourt, with a slight sneer. "Why not?" said Deronda, covering himself.[64]

Grandcourt's sneer denigrates anyone who would offer too much respect to animals—an unsurprising position from a cold-hearted character already shown to be an animal abuser. But his gibe also calls attention to the reality that living, breathing animals linger around these characters, hidden beneath the hypnotic prose of Eliot's scenic description. The response of the warmhearted, open-minded Deronda invites the idea that such recognition could be more than just a joke. "Why not?" he responds, "covering himself." Deronda literally covers himself by restoring his hat to its normal position on top of his head. But he also metaphorically covers himself by holding open the possibility that these animals might be worthy of his respect—and by suggesting that he might be open to paying it to them.

With her strategic use of scenery Eliot covers herself, too. Refusing animals and other creatures the preferential status of characters in her novels, she nevertheless registers their continued presence in the background. Both there and not there, the nonhuman beings in Eliot's social novels hint at the reality of a more extended interspecies community that her work cannot afford to explore—not yet. The slow, self-conscious process by which Eliot excluded such creatures serves as a reminder that these exclusions are neither inevitable nor irreversible. The human focus of her social novels was, finally, a formal choice. It reflects not the inherent anthropocentrism of novels or storytelling but the practical importance of turning readerly attention elsewhere for the time being.[65] Her work calls up the possibility of a less human, more ecological mode of storytelling that does not sacrifice such expansive ethics, even if those stories have to wait for some unascertained future.

As it happened, that future was already on the horizon. While Eliot worked on *The Mill on the Floss,* Charles Darwin overcame decades of reticence and unveiled his revolutionary understanding of biotic relationships in *On the Origin of Species.* Eliot and Lewes actually read the book together when it first came out in November 1859, but Eliot did not grasp its novelty or revolutionary significance at the time.[66] While decidedly unscientific in the strictest sense of the word, Darwin's unified theories of nature effectively spelled the end for natural history, which would soon be replaced by the more rigorous investigations of evolutionary biology and the environmental sciences. Within a few years, Darwin's application of Malthusian storytelling to the entirety of the natural world led one admirer, the German naturalist Ernst Haeckel, to coin the word *Oekologie* to describe this new approach to studying interspecies relationships. It led another admirer, an English architect named Thomas Hardy, to try his hand at writing social novels that reintroduced the natural world into our imagination of what it meant to be human. But given the tools at his disposal, that task would be much harder than he imagined.

CHAPTER 4

The Story of Ecology

On September 14, 1895, the writer George Gissing boarded a train to Dorset to spend a weekend in the country. The holiday was an indication of Gissing's newly solid footing as a novelist. After financial struggles at the start of his career—including prison time for theft, difficulty supporting his chronically ill wife, and multiple failures to find a publisher for his books—Gissing hit his professional stride in the 1890s. He was finally receiving respectable press and respectable pay for his brutally honest depictions of urban poverty. His escape from the vicinity of London was thus a healthy sign of his improved financial situation. It was also a mark of his ascension to a kind of literary elite: on this trip, Gissing would be staying with one of the most prominent authors of the era, Thomas Hardy.[1]

Hardy and Gissing had met before. They read each other's work, and they shared a broadly pessimistic outlook on life.[2] Nevertheless, they made something of an odd couple. There was, first, a discrepancy between the two in age and professional standing. At thirty-seven Gissing was not exactly young. Still, Hardy was nearly twenty years his senior, and the older man had long enjoyed the kind of critical and financial success Gissing was just beginning to taste. Furthermore, Gissing's success was tied to his detailed portrayals of a world utterly absent from Hardy's fiction: the slums and tenements of London. Hardy, by contrast, was famous for his attention to the English landscape. He was widely revered for the stunning natural settings of his tragedies, landscapes that drew on his childhood in England's West Country. If nothing else, then, Gissing's trip would include an intriguing change of scenery. It held the promise of a new intimacy with the natural settings so widely admired in Hardy's work—landscapes that would be all the more absorbing with Hardy himself on hand to act as resident naturalist and guide.

Like so much in Gissing's life, however, the Dorset trip was destined to end in disappointment. Before the weekend was over, Gissing realized that Hardy's extraordinary depictions of nature were not underwritten by expert knowledge of the local flora and fauna. "I find that he does not know the flowers of the field!" the crestfallen Gissing observed in a letter to his brother.[3] Nor was Gissing alone in his disillusionment. To this day, Hardy's writing convinces readers that the author must have been an experienced natural historian, an inveterate rambler of the open country. His early fame rested on his remarkable powers as "a landscape novelist." Since the rise of environmentalism in the 1970s this reputation has only expanded, with a growing chorus of critics redoubling their admiration for Hardy's portrayal of the natural world. Many now champion him as a prescient environmental thinker—perhaps the first truly ecological novelist.[4] But this obstinate belief that Hardy was closely attuned to the natural world does not reflect the realities of Hardy's life or his learning. As his biographer Michael Millgate notes, Hardy "seems in general to have spent a good deal less time outdoors than either his birth, his vocation, or his choice of domicile might have suggested. People meeting him for the first time often remarked upon the paleness of his complexion and his failure to look like a countryman."[5]

The dismay of visitors like Gissing is significant. It highlights the growing fissure that separated realist fiction from the natural sciences toward the end of the nineteenth century. By the 1890s it was possible for a writer like Hardy to produce depictions of the nonhuman environment that struck fiction readers as dazzling displays of knowledge about plants and animals—and it hardly mattered that the nature depicted was uninformed by the emerging science of ecology. The mere act of including natural surroundings in the novel's imagination of community set Hardy apart from his peers, writers who more or less confined themselves to the more familiar world of human social relations. The recent resurgence of critics celebrating Hardy as a model of ecological authorship suggests that little has changed. We still hail the mere introduction of nonhuman nature into social storytelling as if it were radical, insightful, and inherently ecological. It is no longer possible to visit Hardy in person to uncover the gaps in his knowledge of the nature that surrounded him. But that sort of visit is unnecessary, because Hardy's tendency to misconstrue ecology is written right into the pages of his novels. It becomes apparent as soon as his depictions of the land are read alongside the writings of contemporary scientific investigators, particularly the accounts of the naturalist long counted among Hardy's foremost influences: Charles Darwin.[6]

As it happens, both Hardy and Darwin were fascinated by one of the most typical landscapes of Britain—the bare, windswept stretches of thorns and heather commonly known as heathlands. In the nineteenth century, unbroken expanses of heathland covered tens of thousands of acres of southern England. For most observers, their wild and unpeopled appearance conjured the sense of a primitive, inhuman world far removed from the onslaughts of modernity. Hardy was no exception. He had been born in a cottage built of mud and straw perched on the outskirts of one such heath. He used an imagined version of his birthplace as the setting for *The Return of the Native* (1878), a pivotal work in his turn away from romance and toward more realist stories, tragic tales of human lives crushed between the interlocking gears of heartless natural necessity and senseless social conventions.[7] Hardy's native knowledge enabled him to evoke the sublime monotony of such landscapes with an unsurpassed vividness. Nevertheless, his widely admired depictions of Egdon Heath were already out of step with emerging scientific understandings of heathland ecosystems. Almost two decades earlier, Darwin had included English heaths in *On the Origin of Species* as a prime example of the mutualistic relations of humanity and the natural world. Darwin's writing revealed how these apparently primordial, untamed landscapes were in fact man-made places, the extraordinary outgrowths of millennia of reciprocal, interdependent exchanges between the humans, plants, and animals that relied on the heathlands' carefully maintained food webs for their shared survival.

The difference between these understandings of heathlands—one primitive, timeless, and untouched, the other man-made and profoundly human—was not a simple matter of literary ignorance confronting scientific expertise. It was, instead, the product of divergent narrative forms. Darwin drew on ecological plotting techniques to reveal how different species came together to produce this seemingly static, timeless, depopulated landscape. As one of the intellectual kindred of Malthus and Martineau, Darwin sought to produce careful accounts of how matter and energy changed hands through physical exchanges—even if those exchanges crossed traditional boundaries between nature and society, the human and the nonhuman. As an heir to the increasingly anthropocentric traditions of literary realism, Hardy took a different approach. He was schooled in the bland scenic description of Eliot and others but unreconciled to it. In his attempts to push back against the literary exclusion of the nonhuman, Hardy reconfigured realist approaches to setting, imbuing the land with animation and transforming it into a living creature, a vibrant embodiment of the totality of nonhuman nature.[8] The resulting picture of Egdon Heath was innovative and unforgettable. It would go on to shape much of Hardy's later writing about humanity's relationship to the natural world. It would also contribute to the larger public understanding—or really, misunderstanding—of heathlands for decades to come. It was a misunderstanding that brought disastrous consequences in its wake.

Captivated by the enduring image of heathlands as primordial wastes and obstructions to human civilization, twentieth-century policymakers and landowners wrote off such places as unproductive and effectively worthless. In the hundred years following the publication of *The Return of the Native*, almost all of England's lowland heaths were eradicated. They fell victim to general neglect exacerbated by the economic might of foresters, developers, and government officials who targeted such ecosystems for so-called improvement. It was only after the widespread disappearance of lowland heaths that environmentalists realized how eliminating heathland had also eliminated the forms of life that depended on it, human and otherwise.

The fate of English heathlands makes for a dispiriting chapter of Western environmental history. But there is value, too, in retracing the erasure of these landscapes in reality and in fiction. It clarifies how far even the most elaborate fictional depictions of nature had departed from ecological and economic understandings by the end of the nineteenth century. It emphasizes how the failure to unify such discrepant mindsets can produce awful, irreversible consequences. And it underscores the importance of literary structures to all these specialized discourses, demonstrating the enduring need to pay attention to the forms of the stories we tell if we want to develop more accurate, more ethical ways of accounting for humanity's relationship to the natural world.

CHARACTERIZING NATURE IN THOMAS HARDY

Looking back over his own prodigious output in preparation for the 1912 Wessex Edition of his fiction, Thomas Hardy decided that many of his best-loved works could be summed up under a single heading: "Novels of Character and Environment."[9] In a time of escalating environmental crises, it's tempting to see this designation as an acknowledgment of ecological relations being central to his fiction. When Hardy coined the category, however, *environment* was still a relatively vague word, one loosely applied to any kind of spatial location, context, or surrounding circumstance. The term *environmentalism* had a similarly alien meaning in Hardy's lifetime, one that arose from this more abstract sense of environment: *environmentalism* named the belief that an individual's surroundings or context played a determining role in their life and characteristics.[10] It was not until the century following Hardy's death that a more specific meaning, aligning *the environment* with nonhuman nature, gained political currency. With the rise of conservation movements after World War II, *the environment* entered common use as a way of referring to the natural or nonhuman world. *Environmentalism* soon became a catchall phrase for any political or ethical movement advocating not just for human groups but for the planet.[11] The environment Hardy believed he was writing about, in short, was of a much broader and vaguer kind than our own. He would not have associated the term *environment* primarily (much less exclusively) with the natural settings that still make his novels so striking today. It is wise, in other words, to assume a cautious approach to any environmental insights supposedly invested in his settings.

Of all those settings, the landscape of Hardy's midcareer novel *The Return of the Native* is probably the most famous. What makes his account of Egdon Heath so extraordinary is Hardy's immediate and total departure from the conventions of landscape description. Eliot and other mid-to-late Victorian realists typically drained plants and animals of the fascination that transforms human beings into dynamic characters. From the very first pages of *The Return of the Native*, Hardy pushes back against the monotonous stasis of realist scenery. There he treats Egdon Heath not as a bland backdrop but as a being worthy of the same kinds of attention and care most novelists assign to a dynamic, animated character. The idea sounds alluringly ecological—but the story proves otherwise.

Technically, *The Return of the Native* is not the story of Egdon Heath at all. It focuses on Clym Yeobright, one of the most promising young people ever born and raised on the sprawling waste of Egdon. After years of professional success as a jeweler in both England and France, Clym unexpectedly returns home to take up residence on the stark, solitary landscape of his youth. As he explains to his baffled family and fellow heath-dwellers, Clym has made a conscientious decision to abandon the superficial pleasures of modern human society. His new plan is to open a schoolhouse dedicated to educating and uplifting the locals, who are widely and somewhat disparagingly referred to as "heath folk." As this is a Thomas Hardy novel, Clym's plan eventually ends in tragedy. A disastrous marriage and unanticipated health issues soon combine to shatter Clym's hopes and upend the lives of everyone attached to him.

Yet this technical summary of the novel does not do it justice. Since *The Return of the Native* first appeared, readers have noted that Clym is not really the central figure he is supposed to be. His story is effectively subordinated to its setting, while Egdon is elevated to the status of a character in its own right.[12] Clym does not even appear on the scene until the second book, roughly a third of the way through the novel. Before that, the heath and the customs of those living on it occupy center stage. The book opens with a chapter whose title is a riddle: it promises to unveil "A Face on Which Time Makes But Little Impression."[13] The puzzling phrase combines the individual and human idea of a face with the inhuman characteristic of agelessness—total immunity from the ravages of history. The answer to the riddle turns out to be Egdon Heath itself, the primordial landscape whose superficial "face" remains unchanged by the fickle trends of human society.

This initial tension between humanity and inhumanity will come to be the setting's defining feature. At first there is something almost comforting in Egdon's personification. It suggests a harmonious relationship between humanity and the nonhuman. This comforting harmony is reinforced by the many correspondences the text establishes between human bodies and the terrain. It is hard not to delight in passages that describe how an ancient road traversing the heath "bisect[s] that vast dark surface like the parting-line on a head of raven hair" (13) or how "a knot of stunted hollies" creates a black spot that "in the general darkness of the scene [stands] as the pupil in a black eye" (261). The writing is lovely. Taken together, these anthropomorphic passages appear to validate Hardy's early characterization of Egdon as "perfectly accordant with man's nature—neither ghastly, hateful, nor ugly: neither commonplace, unmeaning, nor tame; but, like man, slighted and enduring" (11).[14] Harmonious as they seem, however, these resonances between humanity and landscape are purely metaphorical. They are imaginative images that shape the tone and ambience of the text, but they say nothing about material relationships between species—the kind of relationships that provide the raw material for understanding an ecosystem and our place within it. To understand why Hardy's metaphors never develop into ecological insight, it is useful to recall the difference between metaphors like Hardy's and the metonyms that are so fundamental to realist storytelling.

Metaphors derive their power from the way they link together two distinct, unrelated things. All the relationships Hardy describes between heathlands and human beings engage in this kind of metaphorical yoking together of otherwise unrelated objects. *The Return of the Native* equates the heathland to a human face, scraggly furze to a neatly styled coiffure, and holly trees to an open eye. By bringing such unrelated things or images together, metaphors propose stunning artistic and conceptual unities. The liveliness of a metaphor, however, depends on its freshness. It is poetic and thought-provoking only so long as it proposes an unexpected, unsubstantiated relationship. A metaphor loses its power if it is ever widely adopted or proves to equate two things that are actually materially related. At that point the association it creates becomes so commonplace that people no longer recognize it as an association at all, and they overlook its symbolic, poetic aspects. The result is known as a dead metaphor. So, for example, the idea of a clock's face or a flower bed might once have been strikingly poetic—a linking of two unrelated things. By now, however, calling a clock's dial its face and a planted area a bed has become so widespread that the phrases no longer have any metaphorical power. They have become standard names for things they once poetically described.

The process that leads to dead metaphors was well understood even in the nineteenth century. "Language is fossil poetry," the poet and essayist Ralph Waldo Emerson observed in 1844. "As the limestone of the continent consists of infinite masses of the shells of animalcules, so language is made up of images, or tropes, which now, in their secondary use, have long since ceased to remind us of their poetic origin."[15] Often, the puns and other wordplay so beloved of Victorian nonsense authors amuse us because they reanimate dead metaphors, making us see language differently. Lewis Carroll's *Through the Looking-Glass* (1871), for example, features both a clock with an actual face and flowers sleeping soundly in their beds until they are awoken by Alice wandering through the topsy-turvy looking-glass world.[16]

When Hardy uses metaphors, then, he establishes lyrically powerful equivalences between humanity and the landscape. But those equivalences get their vividness from the conviction that humans and heathlands are essentially distinct, dissimilar, and even alien to one another. Ecological relationships, by contrast, are not metaphorical at all. They are based on actual physical exchanges between living and nonliving things related in space and time—exchanges where one body's matter and energy are transferred to or replace another body's. There is a literary term for this sort of equivalence or association, too. Whereas metaphors use language to equate two distinct, conventionally unrelated things, metonyms use language to build an association between two things closely related in space and time. Referring to the queen as *the crown,* or calling reporters *the press,* are classic examples of metonyms. In each case, one thing (the queen, the reporters) is being defined by its relation to something closely linked to it in space and time (the crown, the printing press). As the linguist Roman Jakobson famously observed, metaphors and metonyms are essentially opposed to each other, and realist storytelling is an inherently metonymic act. It traces relationships as they unfold between beings connected in space and time, moving from one to the next to drive home those material connections.[17]

Ecological storytelling is a form of realism, and it is profoundly metonymic. It focuses on how each being's fate depends on—and is to some degree representable through and interchangeable with—the fate of other creatures related to it in space and time. Hardy's imagination of the relations between humans and nature, by contrast, is profoundly metaphoric. Entranced by the obvious differences between humans and heathlands, he uses that tension to power a series of lyrical images, images that find imaginative resonances between these distinct and opposed things. The poetic and metaphoric fascination of such resonances would be eroded if humans and heathland turned out to be related to or dependent on one another. And so, for all the apparent juxtapositions between humans and heathland that occur over the course of the novel, Hardy maintains the two as distinct and oppositional categories. Character and environment are such an important pairing in Hardy not because he sutures them back together but because their tension powers the engine that drives his whole artistic process.

Nowhere is this clearer than on Egdon Heath. While critics and readers may rhapsodize over the landscape's characterization, the setting only comes alive through a kind of elaborate imaginative sleight of hand. As Hardy himself admits, the heathland is an almost spectacularly flat, static, and monotonous setting. He insists, after all, that the heath "had been from prehistoric times as unaltered as the stars overhead. . . . The sea changed, the fields changed, the rivers, the villages, and the people changed, yet Egdon remained" (11–12). In order to animate the overwhelming stasis of this landscape, Hardy zeroes in on the very features of Egdon that seem most incompatible with the scale of human life—its flatness, its largeness, its stillness, and its darkness—and he transforms them through acts of artistic imagination.

Under the captivating spell of Hardy's prose, everything inhuman about Egdon begins to appear as the expression of a human personality. The very lack of activity visible on Egdon becomes a sign, for those savvy enough to read it, of a buried life of deep and inscrutable animation. Where others see inertness, Hardy argues, a true intimate of the heath will see patience and profundity—a place "full of a watchful intentness" (9). This revision of setting does not so much find activity in the landscape as fancy that there may be something deeply impressive about just how inactive Egdon is. Its changelessness is recast as a concerted effort to suppress its own internal life, a life that plays out over stretches of time hard for humans to fathom: "This was not the repose of actual stagnation, but the apparent repose of incredible slowness. A condition of healthy life so nearly resembling the torpor of death is a noticeable thing of its sort; to exhibit the inertness of the desert, and at the same time to be exercising powers akin to those of the meadow, and even of the forest, awakened in those who thought of it the attentiveness usually engendered by understatement and reserve" (16). Hardy's decision to render dynamic, enticing, and human what first appear as the most boring traits of the landscape yields impressive results. Egdon emerges from *The Return of the Native* looking like a highly distinctive character, one whose physical attributes become active influences on individual human lives.

There is a duplicity behind this dynamism, however. Hardy has not actually revealed anything lively or engaging in the heath; instead, he has rewritten the monotonous stasis he sees as a kind of wise, willful resistance to change. So, viewed through Hardy's metaphoric reframing, Egdon's uniformly dark coloration is not a natural and unchanging feature of the landscape. It is, instead, a symbol of the land's dogged refusal to participate in daily cycles of light and darkness. "The face of the heath," Hardy writes, "by its mere complexion added half-an-hour to eve: it could in like manner retard the dawn, sadden noon, anticipate the frowning of storms scarcely generated, and intensify the opacity of a moonless midnight to a cause of shaking and dread" (9). It's no wonder generations of readers have found such writing unforgettable and awe-inspiring. The verbs here are astonishingly active and engaging for a landscape—the land "add[s]," "retard[s]," "sadden[s]," "anticipate[s]," and "intensif[ies]." But all this dynamic writing is finally reducible to rhetoric, to linguistic conjuring tricks. It is a poetic set of phrases describing how a dark and stable landscape continues to appear dark and stable under different atmospheric conditions. The end result does not discover any material liveliness in the actual environment, only in its creative characterization.

The fascination of these techniques comes at a cost, too. In order to recast the unchanging features of the setting as willful responses to human activity, Hardy must interpret the land's inertia as intentional standoffishness—an obstinacy that sometimes explodes into rebellion against humanity. If Egdon Heath appears passive, Hardy rather paradoxically suggests, that is only because the landscape is busy opposing all hopeful, willful human action. It does not simply sit apart from human agents; rather, it perversely works to "intensify . . . [the] shaking and dread" of anyone who encounters it (9). Instead of seeming to exist separate from human life, the novelistic environment now appears to participate in it—but only as a conscientious objector.

If Egdon opposes itself to daily changes of light and dark, its opposition does not stop there. It opposes historical changes as well as daily ones, something Hardy uncovers through his consultation of the real, eight-hundred-year-old British territorial survey known as the Domesday Book. "This obscure, obsolete, superseded country figures in Domesday," Hardy notes. "Its condition is recorded therein as that of heathy, furzy, briary wilderness,—'Bruaria.' Then follows the length and breadth in leagues; and, though some uncertainty exists as to the exact extent of this ancient lineal measure, it appears from the figures that the area of Egdon down to the present day has but little diminished. 'Turbaria Bruaria'—the right of cutting heath-turf—occurs in charters relating to the district. 'Overgrown with heth and mosse,' says Leland of the same dark sweep of country" (11).

Extending his art of remaking stasis into something dynamic and relational, Hardy reads the heath's permanence as a pointed mockery of human culture. "Ever since the beginning of vegetation its soil had worn the same antique brown dress, the natural and invariable garment of the formation," he explains. "In its venerable one coat lay a certain vein of satire on human vanity in clothes" (11). By the end of this introduction, Egdon exhibits a smoldering resentment toward society: "Here at least were intelligible facts regarding landscape. . . . The untameable, Ishmaelitish thing that Egdon now was it always had been. Civilization was its enemy" (11).

In short, there is nothing consoling about the relations between humanity and nature that Hardy's literary eye picks out. His metaphors may discover among the rolling curves of scrub a familiar-feeling face, eye, or shoulder; he

may hear in the wind something akin to a voice. Taken as a whole, however, this personified landscape turns out to be anything but friendly to the people who eke out an existence upon it. Inherently opposed to the human, Egdon Heath seethes with anger toward humans and their history. Its apparent passivity is merely a sign that it "await[s] one last crisis—the final Overthrow" (10). If heath-dwellers try to cultivate a relationship with such a place by, say, converting a portion of it to life-sustaining farmland, Egdon wages war against them—a war of attrition in which Egdon always emerges victorious. The native referenced in the novel's title, Clym himself, observes the scars of these wars when he returns from Paris. Surveying the landscape, he mentally remarks on "the attempts at reclamation from the waste, [where] tillage, after holding on for a year or two, had receded again in despair, the ferns and furze-tufts stubbornly reasserting themselves" (172).[18]

Who could feel at home in such an environment? Only misanthropes, as it turns out. When characters like Clym embrace life on the heathland, they do so because they share the land's barely veiled hatred of humanity. Clym's desire to return to the heath is driven by a sort of wistful recognition of the heath as deeply antisocial—and as therefore opposed to the fashionable world that he, too, has come to despise. Although he is occasionally romanticized as being "inwoven with the heath," Clym's upbringing did not immediately breed in him any love for the landscape (166). By his own account, he concluded early on that "this place was not worth troubling about" and that "life here was contemptible," a position that led him to take a job as a jeweler in the seaside resort of Budmouth and eventually in Paris (168). His return to Egdon represents a belated, ambivalent endorsement of this "obscure, removed spot" as the polar opposite of "the French capital—the centre and vortex of the fashionable world" (109). So when he returns and sees ferns and furze overrunning the lands that formerly belonged to (failed) farmers, Clym responds with sympathy for the heath and disgust toward his fellow human beings. "[H]e could not help indulging in a barbarous satisfaction" at the heath's victory and humanity's loss, Hardy admits (172).

When Clym's relationship with heathland is understood in this light, the novel's otherwise unconvincing marriage plot starts to make sense. Not long after he returns, Clym begins a flirtation with the dark, brooding Eustacia Vye, who is not a native of the place but is a longtime resident. Unlike Clym, Eustacia hates the heath and wants nothing more than to escape to the very civilization Clym has left behind. She longs for the rich, superficial society of seaside resorts and city living. Clym delights in the stoic solitude of nature on the heathland. Yet different as they are, both characters at least agree in their understanding of what Egdon is. Like Clym, Eustacia sees the heath as a sprawling, agentic enemy thwarting the civilization she craves.

The polarized, magnetic attraction of Clym and Eustacia illustrates the essentially complementary relation between Clym's love and Eustacia's hate. Their positions are as unstable as they are extreme. As a result, both characters vacillate wildly between identifying with their inhuman environment and being horrified by it. Eustacia's repugnance toward the heath, for example, masks her extraordinary similarity to it: over time, she has acquired the suppressed rancor, contrariness, and isolated majesty that—to the eyes of a newly converted antimodern like Clym—align her with the environment she despises. "Egdon was her Hades," Hardy remarks, "and since coming there she had imbibed much of what was dark in its tone, though inwardly and eternally unreconciled thereto" (67). Among other things, she drinks in the land's misanthropy, which distances her from other people. "I have not much love for my fellow creatures," she confesses to a bewitched, bewildered Clym. "Sometimes I quite hate them" (183).

Clym shows the same unstable combination of identity with and repulsion toward the environment he inhabits. For all his supposed intimacy with Egdon, he experiences several disorienting episodes when he realizes that the landscape's aggressive indifference to humanity applies to him, too. After Eustacia accepts his marriage proposal, for example, Clym is unsettled by the absence of feeling in the surrounding scenery: "As he watched, the dead flat of the scenery overpowered him. . . . There was something in its oppressive horizontality which . . . gave him a sense of bare equality with and no superiority to a single living thing under the sun" (203–4). Clym confronts the inhumanity of his natural surroundings one more time late in the novel, when a loved one dies of exposure in the very same landscape. Stepping out of the shack where the body lies, Clym finds himself face-to-face with "the imperturbable countenance of the heath, which, having defied the cataclysmal onsets of centuries, reduced to insignificance by its seamed and antique features the wildest turmoil of a single man" (312). These brief, revelatory glimpses of the heath's horror draw on both its characterization and its inhumanity. In them, Hardy lingers over the hypnotic repulsion spawned by this setting's internal dissonance, its impossible combination of personification and impersonality.

There is a name for this eerie emotional terrain, this place where growing sympathy and identification with a nonhuman object suddenly reverse themselves and give way to revulsion, alienation, and horror. Designers and roboticists call it "the uncanny valley," a sort of aesthetic and emotional Bermuda Triangle that should be avoided at all costs. As the engineer Masahiro Mori first explained in 1970, we human beings are drawn to our own features. We typically respond positively to anthropomorphic appearances in nonhuman objects, delighted to recognize traces of ourselves where we least expect them. But if objects become very humanlike, they start to become unclassifiable. They are at once too human to be objects and too inanimate to count as truly human. In this murky philosophical territory, the humanity of nonhuman things goes from delightful to disturbing. The humanized objects confound our sense of what constitutes the human in the first place. As the theorist Sianne Ngai notes, the discomfort such objects produce is tied to our subconscious realization that "if things can be personified, persons can be made things."[19] This unsettling tension occurs whenever a thoroughgoing personification is applied to something that continues to defy human standards of personhood. In such cases, the inhuman plays a creepy sort of peekaboo with the viewer, as the personification alternately masks, and then throws into stark relief, the object's violation of our expectations of humanity.

With his experimental reconfigurations of character and environment in *The Return of the Native,* Hardy inadvertently stumbles into the uncanny valley, and he maps it with all the fervor of a pioneer. Like a corpse or a glassy-eyed talking doll, Egdon Heath produces this vehement and highly recognizable response in observers—including characters like Clym and Eustacia. Because it is so insistently personified, Egdon kindles some sympathetic recognition in readers and other characters. But this sympathy unpredictably gives way to revulsion and horror whenever the narrator or characters recall that this sweep of nature is inherently inhuman. These polarized emotional patterns are on vivid display in *The Return of the Native,* but they are also typical of Hardy's later work. Indeed, critics have long recognized an extreme, unstable alternation between attraction and repulsion toward nature as a hallmark of Hardy's best writing. In her groundbreaking study of Darwinian influences on Victorian literature, Gillian Beer dubbed this quirk of Hardy's his "creative vacillation" between "[h]appiness and hap."[20] George Levine describes it more simply as "Hardy's ambivalent relation to the natural world."[21] Whatever the preferred terms, the pattern is unmistakable.

Egdon Heath shows how that pattern emerges as the by-product of a certain way of representing the natural world. It was an approach that

Hardy pursued for the rest of his career. Although none of his later novels personify the landscape to the same extent as *The Return of the Native,* they indulge in some personification of nature—and they express despair about the outsize power this inhuman figure exerts over the human individuals entangled with it. So, for example, *Tess of the d'Urbervilles* (1891) features an entire cast of characters confused and deceived by "the vulpine slyness of Dame Nature."[22] Dame Nature is essentially a procuress, a heartless personification of earthly life whose job is to implant sexual desire into otherwise innocent creatures to ensure that they reproduce. Her ministrations leave human beings "writh[ing] feverishly under the oppressiveness of an emotion thrust on them by cruel Nature's law—an emotion which they had neither expected nor desired."[23] In *Jude the Obscure* (1894–95), this already personified force of nature becomes almost cartoonishly human. Biological desires take the shape of a brawny arm that bullies the hero Jude Fawley, clutching and dragging him down a path toward sexual partners and personal destruction entirely against his will: "In short, as if materially, a compelling arm of extraordinary muscular power seized hold of him. . . . This seemed to care little for his reason and his will, nothing for his so-called elevated intentions, and moved him along, as a violent schoolmaster . . . towards the embrace of a woman."[24]

Despite the attraction of these uncanny characterizations of nature, they are not ecological. They ascribe certain human traits and powers to the natural world, but they continue to draw a sharp line between this sprawling, agentic nonhuman nature and the human beings forced to engage with it. As Hardy's individual characters struggle against this weirdly humanoid aggregate of all things inhuman, they are overwhelmed and overpowered—the natural result of taking on such an oversized foe. And it is, finally, only as a foe that Hardy comes to understand nature, despite his knowledge that we are bound up in it. His formal strategy of gathering vast swaths of land and every nonhuman living on them into a unified character—and later, of bundling all life forces into a single, monstrous personification of nature—leads to captivating visions of Nature as a toweringly nonhuman person. This Nature diminishes to insignificance all human individuals, whom Hardy treats as paradoxically distinct from and inextricably caught up in it.

The result is unforgettable literature, and it has inspired generations of readers to direct their thoughts and feelings toward the natural world. That may be a positive development in and of itself. But Hardy's style of encouraging environmental feeling has a dark side, and his approach bears only a passing resemblance to ecological thought. His later writing is marked by a deep suspicion and ambivalence toward the environment, a volatile mixture of admiration for nature's anthropomorphic aspects and repulsion at its inherent inhumanity. Fascinating as Hardy's uncanny nature can be, it distracts from the circuits of material relations that connect humans and nonhumans, the circuits Aldo Leopold would later identify as the basis of ecological understanding.[25] An ecological approach to such interconnections reveals a nonhuman world that is more hopeful than Hardy ever imagined—and more imperiled, too.

CHARLES DARWIN'S PLOTS OF LAND

Hardy's writings on heathland have secured him a privileged place in the annals of English environmentalism. Conservationists celebrate him as a contrarian who broke with convention to recognize the value of heath landscapes that were widely despised at the time and have since been all but eradicated. In *The History of the Countryside* (1986), Oliver Rackham's magisterial survey of British environmental history, Rackham praises Hardy for offering an important counterpoint to "ericophobia"—Rackham's word for the hatred of heathland that became commonplace in the late nineteenth and early twentieth centuries. Rackham groups Hardy with a small cadre of writers who captured "the glory and mystery and freedom of the heath" before adding, Cassandra-like: "But few listened."[26] The National Trust's book-length tribute to English heathland follows Rackham's lead, remarking that "it was Thomas Hardy who was to become heathland's greatest literary champion."[27] The geographer Christopher Tilley has gone so far as to dub Hardy the "principal 'historian' and 'ethnographer' of the English lowland heathland," despite the fictional status of his landscapes.[28]

As it turns out, though, Hardy got heathland wrong. The landscape he singles out in *The Return of the Native* as a "great inviolate place" possessing "an ancient permanence which the sea cannot claim" was entirely manmade (12). England's heathlands are inherently unstable ecosystems, communities dependent on human folkways of grazing, cutting, and burning to persist. In one sense Hardy was an adept chronicler of the details of life on the heath; these acts of grazing, cutting, and burning are described throughout the novel. Nevertheless, Hardy could not imagine a significant metonymic connection between the human activities he recorded and the environment in which they took place. In the novelist's vision, every inroad human effort seemed to make toward shaping or modifying the ways of nature could only conclude with a dead end. Despite the novelty of his literary techniques, he could see no material bridge across the gap separating human characters from nonhuman settings, the distance that came to typify the work of realists like George Eliot. Yet where Hardy failed to recognize ecological interconnection, one of his most important forebears had already succeeded. Twenty years prior to Hardy's writing, Darwin concisely laid out a theory of the human origins of heathland in "The Struggle for Existence," chapter 3 of *On the Origin of Species*. There, Darwin turned to heaths as an illuminating study in "the mutual relations of all organic beings"—the study that Haeckel touted as a new branch of science named *ecology*.[29]

Returning to Darwin's narration of heathland communities helps explain why he saw the interspecies connections that eluded Hardy. It suggests that ecological storytelling is not simply a matter of creating narratives in which human and nonhuman beings appear to interact. Rather, ecological storytelling involves finding the best literary forms to accurately capture the effects of such interactions—a process that involves viewing the world from multiple perspectives and scales. Only then is it possible to determine which agents and actions matter and how those actions ripple across scales to shape the overarching ecosystem. The ability to elaborate such effects depends, in short, on judicious use of that narrative assignment of agency known as characterization. Careful characterization makes it possible to see how apparently local, individual actions may have effects that are distributed widely across space and time. Darwin's hypotheses on heathland demonstrate how the scaling of character proves key to ecological understanding. They also show how a fixation on landscape can impede such understanding, making thinking across scales all but impossible.

At first, Darwin's encounter with heathland in the *Origin* closely resembles Hardy's later imagination of Egdon Heath. "[O]n the estate of a relation" in Staffordshire, Darwin recalls, he once found himself amidst "a large and extremely barren heath, which had never been touched by the hand of man" (59). Surveying the scene, he was surprised to find that the landowners had successfully converted some of this apparently timeless, treeless landscape into a more economically productive type of terrain: "[S]everal hundred acres of exactly the same nature had been enclosed twenty-five years previously and planted with Scotch fir. The change in the native vegetation . . . was most remarkable" (59–60). There was, he reports, a radical shift in "vegetation," "insects," and "insectivorous birds" over the

course of that quarter century, a transformation that hinted at the surprising mutability of a landscape that looked permanent (60).

A second, subsequent experience with heathland made Darwin realize just how little effort it took to destabilize such places. In Surrey, he found yet another landowner who appeared to be fencing off heathlands to turn them into fir plantations. As he looked closer, however, it dawned on the naturalist that the firs were not planted in any orderly fashion; they were "springing up in multitudes, so close together that all cannot live" (60). Darwin realized these were "self-sown firs," not products of direct human intervention (60). Recognizing that he had misunderstood the place and the changes overtaking it, he began searching for new perspectives that would clarify why every fenced-off section sported a dense stand of evergreen trees. "When I ascertained that these young trees had not been sown or planted," he recounts, "I was so much surprised at their numbers that I went to several points of view, whence I could examine hundreds of acres of the unenclosed heath" (60). His attempts to see the scene anew proved useless, however: each vista only disclosed the same monotonous stretch of furze and heather that Hardy would render so memorably. On that unenclosed landscape, Darwin recalls, "literally I could not see a single Scotch fir" (60).

After failing to make sense of the heath from assorted scenic viewpoints, Darwin did something remarkable: he literally pulled the landscape apart. Shifting his gaze from the horizon to the plants and animals under his nose, Darwin bent over and separated the monotonous stalks of heather to peer beneath them. There he uncovered an entire community of pine trees that did not reach his knees—a crowd of heathland inhabitants he had never seen before. Each of these miniscule seedlings, Darwin realized, starred in its own individual story of thwarted survival. "In one square yard . . . I counted thirty-two little trees," he writes, "and one of them, with twenty-six rings of growth, had during many years tried to raise its head above the stems of the heath, and had failed" (60).

Here as in Hardy, literary language matters. When Hardy personifies the entire heath, he leaves the landscape essentially intact. His personification tacitly admits the stability and internal coherence of landscape as a conceptual category. The result is a totalizing understanding of the land and its nonhuman inhabitants as one kind of "environment"—a monolithic, inhuman mass whose intents and actions operate at a scale that shrinks human individuals into irrelevance. Darwin adopts a similarly scenic perspective during his first investigation into heathland. He quickly abandons it, however, when it proves inadequate to explaining the biological phenomena he is seeing. This move is, in fact, characteristic of Darwin's writings on landscape. He tends to begin from the artistic perspective of a single Romantic observer before breaking the scenery into individual organic and inorganic components. "From the visual unity of the observer's momentary perception," observes James Paradis, a scholar of scientific rhetoric, "we move to the conceptual diversity of [Darwin's] accumulating record of fact."[30] But if Darwin finally arrives at scientific clarity, that arrival does not take the shape of a direct leap from gauzy Romantic landscape to hard-nosed scientific objectivity. His scientific understanding comes by way of yet another aesthetic middleman: narrative. Darwin dissects the scenery to discover a world of agentic characters and entangled storylines secreted underneath it.

The divergent understandings of Darwin and Hardy, in other words, are not a matter of differences between science and literature or between fact and fiction. Instead, they illustrate the different insights made available by different artistic methods. Darwin's physical decision to pull the scenery to pieces is also a philosophical decision to reject the sweeping scale and comparative permanence of the heathland—traits central to the very idea of landscape. This aesthetic rejection enables him to discover the tiny pines. Darwin frames their discovery not as a static fact but as a useful clue that will help him unravel the story behind the landscape's superficial stasis. He almost immediately personifies these saplings, recognizing their agency as characters in their own miniaturized storylines. Darwin's decision reinforces recent arguments that anthropomorphic approaches to the nonhuman can help spur ecological understanding and fellow feeling between species.[31] Yet the divergent approaches of Hardy and Darwin also suggest that personification can be beneficial or baneful for ecological understanding. It all depends on how, exactly, personification is applied.

In Darwin's writing, personification operates at precisely those individual scales that Egdon renders insignificant. Not only does Darwin immediately anthropomorphize the tiny pines in the *Origin* but he singles out the representative struggle of "one of them . . . [that] had during many years tried to raise its head . . . and failed" (60). Like all stories, this plot raises a question of causality: Why had the tree failed? The tree sprouted and began to grow, and then—*What happened next?* Rather than raising his eyes to the horizon (as Hardy might) and assigning this tree's failure to the unforgiving nature of the monolithic landscape, Darwin keeps his attention trained on individual agencies. The question of causation leads him to take stock of other characters who might figure in this narrative and to reevaluate the significance of creatures he had previously missed. Cattle, he realizes, must have "perpetually browsed down" the growth of each seedling before any tree could get high enough to overshadow the heather (60). Although Darwin must have observed the occasional grazing animal in his wanderings, cattle did not factor into his sweeping survey of heathland scenery; they did not seem like noteworthy or effective presences in such deserted places. "[T]he heath was so extremely barren and so extensive," he explains, "that no one would ever have imagined that cattle would have so closely and effectually searched it for food" (60). It is, as Darwin himself says, an imaginative failure that prevents scientific understanding. Hard as it may be to imagine scattered cattle browsing down an entire biome's worth of virtually invisible saplings, the cattle did just that—until assorted acts of fencing prevented them. The exclusion of livestock gave the tree seedlings the opportunity they needed to overgrow the dense heathers. Within a few years, these trees starved the surrounding shrubs of sunlight, converting each enclosure into a small wood. "Here we see," Darwin concludes, "that cattle absolutely determine the existence of the Scotch fir," which in turn determines the existence or disappearance of the heathland itself (60).

Darwin leaves unsaid what contemporary readers could easily infer: cattle are human introductions to such landscapes, so these apparently timeless heaths owed their continued existence to human agriculture and pasturage. Decisive confirmation of this hypothesis would have to wait another hundred years. By the mid-twentieth century, paleobotanists had analyzed enough prehistoric pollen samples to conclude that English lowland heaths like those in Surrey and Dorset were man-made landscapes. Heathers rapidly proliferated across Britain, first with the rise of Bronze Age civilizations and then more dramatically with the onset of the Iron Age. Soon heather pollen outnumbered the pollen of tree species in much of Britain, a sign of major deforestation that coincided with increased technological sophistication among early inhabitants of the island. These discoveries underlie the current ecological understanding of English heathlands as unique biotic communities that develop on acidic, nutrient-poor, well-drained soils as a result of specific patterns of human cultivation. As Darwin speculated, an undisturbed heath quickly converts into woodland, typically one dominated by opportunistic species such as Scots pine (*Pinus sylvestris,* which Darwin and other Victorians called "Scotch fir") and European white birch (*Betula pendula*).[32]

Darwin's ecological reading of heathland is so accurate as to seem almost preternatural—but only if we overlook his own painstaking account of the aesthetic tools he used to decode the phenomena he saw. As scholars have long observed, Darwin's revolutionary insights are inextricable from the language he employed to produce and communicate them. The tropes and metaphors of Darwin's works cannot simply be "skimmed off," Beer notes, to reveal pure scientific ideas beneath them.[33] Typically, however, studies of the narratives and metaphors in Darwin's work focus on natural selection, highlighting the ways he personifies nature as an agent in the eons-long family epic that provides the storyline of evolution.[34] That makes it easy to pass over the way Darwin applies personification and narrative techniques on more local scales. There such techniques work "to imagine a world of nonhuman intent and distributed sentience that is far closer to humanity than previously imagined," as the literary critic Devin Griffiths observes.[35] Applying personification and narrative to deep history enabled Darwin to grasp the mechanism of natural selection. Those same techniques applied to smaller timescales and local actors enabled him to grasp ecological dynamics, or—to use Darwin's own words in the *Origin*—"how plants and animals, most remote in the scale of nature, are bound together by a web of complex relations" (61).

Darwin's reading of heathland demonstrates how conventional approaches to landscape and scenery often stand in the way of accurate ecological storytelling. "[T]he face of nature remains uniform for long periods of time," he admits, using language that anticipates Hardy's favored metaphor for Egdon Heath (61). Yet rather than lingering over and reinforcing this image, Darwin raises it only to negate it. He repeatedly dives beneath the collective personification of the natural world to untangle individual stories of struggle, success, and failure. These stories reveal the contingencies behind nature's superficial stability, showing how "the merest trifle would often give the victory to one organic being over another" (61).

This willingness to think across scales to see how apparently insignificant agents could produce outsize changes in a landscape was a lifelong preoccupation of Darwin's. In his final publication, *The Formation of Vegetable Mould through the Action of Worms* (1881), he recapitulates his enduring commitment to storytelling across scales—and he rails against those who lack the imagination to engage in it. As Darwin describes it, his study of worms, which originated with a paper he had presented at the Geological Society of London nearly forty years earlier, illustrates the importance of attending to "small agencies and their accumulated effects." Only through such close attention is it possible to overcome a particularly pernicious, widespread form of imaginative failure, "that inability to sum up the effects of a continually recurrent cause, which has often retarded the progress of science."[36]

Though much has been made of Hardy's debt to Darwin and his supposedly Darwinian habits of attention, *The Return of the Native* suffers from exactly this inability. Hardy cannot see how the actions of individual agents accumulate, reshaping and redefining the environments they inhabit. What makes Hardy's oversight more surprising is that, in terms of documentary details alone, his account of heathland far eclipses Darwin's. Darwin depicts a stripped-down vision of heaths where the only denizens are heather, pine trees, and (in a belated addition) cattle. With that later addendum, Darwin intimates how heathland was traditionally used as pasturage for grazing animals, including (depending on the terrain and the local customs) sheep, ponies, goats, rabbits, and cattle. What Darwin does not acknowledge is that heather and furze also served as invaluable resources for heath-dwellers, people whose folkways revolved around the seasonal collection and processing of the plentiful raw materials the heath offered.

The Return of the Native dutifully records the details of such folkways, providing a sense of heathland culture absent from *On the Origin of Species*. For all its apparent barrenness, heathland provides a steady supply of resources to those who know how to harvest them. The root systems of heather produce excellent turf—topsoil so matted together that it can be cut, dried, and stacked to serve either as slow-burning fuel in the hearth or as construction material for housing. The prickly furze complements turf as a fuel source, burning hotter, brighter, and much more quickly. Both heather and furze have edible components, and both can be fashioned into a variety of household implements ranging from brooms to cutlery. The ferns or bracken that spring up in wetter portions of heathland can be gathered, too. They were once used for bedding material and burned to produce potash, a compound integral to the manufacture of glass and soap. In a traditional heathland community, then, grazing animals might provide meat, dairy, and leather to inhabitants, while the vegetation that fed those animals could be eaten outright or converted into fuel, construction material, wood, and the chemical components of household goods.[37]

If heaths prove more than capable of sustaining human life, humans themselves prove necessary for sustaining heathlands. Environmental historians now recognize these traditional occupations and folkways as important contributors to the stability of such landscapes. Just like the grazing of domesticated animals, the harvesting of natural resources involved cyclically cutting back the plants of the heathland and overturning its soils—actions that disturbed and denuded the vegetation regularly enough to prevent forests from taking over. Historically speaking, the wide range and long duration of heathlands in Britain provide evidence of an indigenous British way of life. Human and nonhuman animals existed in entangled, symbiotic relations with this scrubby, strange vegetation for thousands of years.

The chorus-like group of heath folk who occasionally weigh in on the fates of Clym and Eustacia in *The Return of the Native* demonstrate Hardy's interest in these folkways. The heath folk include "Sam the turfcutter" (48), "Olly Dowden—a woman who lived by making heath brooms, or besoms" (25), and numerous furze-cutters (eventually including Clym), all of whom know how to turn landscape into livelihood. Despite the fact that Hardy recorded many of the occupations necessary to the special symbiosis of heathland, however, it never occurred to him that human beings were working to co-create this landscape that sustained so many species. In his representations of individuals engaging with their environments, Hardy saw only futility—a futility so extensive that it included every plant and animal individuated against the landscape. A "clump of Scotch fir-trees," for example, offers Hardy an occasion to portray the trees' cruel treatment at the hands of an overpowering environment:

> Not a bough in the nine trees which composed the group but was splintered, lopped and distorted by the fierce weather that there held them at its mercy whenever it prevailed. Some were blasted and split as if by lightning, black stains as from fire marking their sides, while the ground at their feet was strewn with dead sticks and heaps of cones blown down in the gales of past years. . . . On the present heated afternoon when no perceptible wind was blowing the trees kept up a perpetual moan which one could hardly believe to be caused by the air. (268)

This passage shows Hardy commiserating with the trees, which is remarkable in its way. But it also shows him utterly incapable of seeing the trees for what they really are—interlopers encroaching on the heath, interlopers that in time will destroy the landscape entirely. Instead Hardy frames them as symbols of nature's uniform cruelty toward individual striving, a cruelty writ large on what he had mistakenly decided was a monolithic, unchanging landscape.

It is tempting to say that Hardy could not see the encroaching forest for the trees. In fact, however, he had the opposite problem: he could not see individual agents like trees because his eye was trained on what seemed to him like the bigger and more significant scenery. Hardy is known for his fascination with sweeps of time and space so vast and permanent that they appear to rob individual lives of significance. This version of Hardy—as a writer trying to bridge what the critic Benjamin Morgan calls the "scalar leaps and disjunctures" between individual human lives and the "deep time" associated with geology and evolution—is well established.[38] His novels obsess over the stony endurance of the planet, what ecocritic Jeffrey Jerome Cohen has described as the "disorienting realization . . . that we are fleeting, that this place supposed to be home is too ancient and enduring for domestication."[39]

The example of Egdon Heath suggests that Hardy's interest in landscapes, planets, and deep time actually worked at cross-purposes with his interest in environmental relations. Simply put, Hardy's obsession with sweeping vistas made him incapable of appreciating the significance of those "small agencies" that were central to Darwinian thought and to the ecological science that sprang from it. Hardy was an impressive observer of the human and nonhuman denizens of ecosystems; Darwin omits many of the entities who collaborate to create heath landscapes, while Hardy meticulously includes them. But Hardy's storylines fail to imagine that such entities actually matter on a grand scale. Darwin, by contrast, is able to understand such entities as agentic characters and to imagine the effects of their interactions. It is in the act of plotting—the storytelling that brings manifold beings together in some causal, metonymic, and meaningful manner—that Darwin surpasses his literary disciple. Such plotting wears away at the superficially intimidating endurance of nature and geology, playing up the way that interconnection enhances agency.

Reading the heathlands of Darwin and Hardy side by side suggests that the seemingly ecological act of treating nature and its landscapes as coherent, organic wholes may cause more harm than good. Not only does it obstruct ecological understanding but it promotes a kind of existential despair. Lively as they are, Hardy's giant agentic environments establish a fundamental incompatibility between the nonhuman world and the scale of individual human lives. This incompatibility only renders nature's liveliness more horrifying: in Hardy, nature reenters human social storytelling as a dangerously overlooked player whose power is ultimately unstoppable.

In the end, his novels animate old oppositions between humans and nonhumans rather than transcending them. By contrast, Darwin's reading of heathland shows how incompatibilities between land and humanity only *seem* self-evident. They are by-products of established but inaccurate habits of dividing the world into active agents and static backgrounds. One way these habits are passed down through the generations is the traditional literary division between character and setting—a division that can have devastating consequences for human beings and the nonhuman communities we occupy, shape, and sometimes eradicate.

CULTIVATING THE ECOLOGICAL IMAGINATION

While Hardy was busy crafting his unforgettable image of Egdon Heath as a timeless, untamed landscape, the heathlands of his native Dorset were disappearing at a rate of 0.6 acres per day. Industrialization, suburban sprawl, and acts of enclosure were putting pressure on traditional folkways associated with furze- and turf-cutting, foraging, and common rights, unpeopling the heaths and leaving growing stretches of terrain uncut and ungrazed. In many places, the birch and conifer forest held in abeyance for hundreds or thousands of years took over. In others, the increasing efficiency of coal- and gasoline-powered transit made once-remote heathlands newly accessible from city centers, inciting real estate developers to buy up this flat, cheap landscape and transform it into home sites for suburban commuters. (Later, during the world wars, privately held heaths were expropriated by the government for similar reasons: they were cheap, they were relatively level, and they had a reputation for uselessness, making them ideal spots for construction—in this case the construction of military bases, munitions sites, and training grounds.) Heaths that were not neglected, built up, or paved over became targets for the Forestry Commission, which was formed as a response to the shortage of timber in Britain during World War I. The Forestry Commission encouraged the wholesale conversion of heathland into pine plantations, a policy of aggressive afforestation that hastened a process of ecological succession that would have occurred anyway. By the mid-twentieth century, Dorset heaths were vanishing at a rate of 2.1 acres per day, more than triple the rates observable during Hardy's lifetime. When environmentalists finally mobilized to save such landscapes in the 1980s, 80 percent of Dorset's historic heathlands had disappeared—including most of Puddletown Heath, which served as the inspiration for Egdon.

The disappearance of this habitat led to devastating population crashes among British species specially adapted to life in heathlands, including the Dartford warbler (*Sylvia undata*), the red-backed shrike (*Lanius collurio*), the sand lizard (*Lacerta agilis*), the slow worm (*Anguis fragilis*—actually a legless lizard), and the natterjack toad (*Epidalea calamita*).[40]

The loss of British heathland is hard to process because it defies the genres or forms of storytelling we commonly use to understand ecological catastrophe.[41] The rhetoric of modern environmentalism tends to take its inspiration from a particularly somber poetic mode known as elegy. Traditional elegies are lyrical verses lamenting a dead individual. The modern environmental elegies that crop up in magazine articles, newspaper stories, and prizewinning books discard verse in favor of prose. Still, they retain the elegy's mournful lyricism, applying it not to dead individuals but to the death of a once thriving, untouched environment at the hands of modern society. If heaths were what Hardy thought—pristine, primordial relics—their disappearance would fit neatly into the elegiac mode of modern environmentalism. The truth, as always, is messier.

The vision of heathland promoted in *The Return of the Native* actually aligns rather neatly with the vision of those human "improvers" who destroyed such places. Egdon Heath is primordial, inhuman, and apparently indomitable. All these features obscure both the social significance of the landscape and its fragility. Instead, they work to further the notion that heaths are barbaric antagonists to civilization—untamed places that would constitute worthy adversaries for the trained forester or modern agronomist. For all its formal novelty, then, Hardy's misinterpretation of this landscape affirms entrenched ideas of heathland as a wild, inimical, and ineradicable landscape. Those ideas would have devastating consequences.

Yet the very language we use to frame ecological change—eradication, devastation—makes it difficult to explain what really happened to English heathland, or to conceive the true meaning of its disappearance. Heathlands, after all, could not be eradicated or devastated by sudden human interference: they were always already altered. The climatic conditions of Britain mean that natural, untouched heathlands can exist only briefly on the island, springing up in disturbed lowland areas for a short time before they are overtaken by woodland. The man-made heathlands that vanished over the last century are ill-suited to the role of an idealized Nature, an innocent, long-suffering maiden threatened by the villainous greed of modern human beings. But heathland's failure to fit comfortably with conventional stories of ecological loss is exactly what makes it such a valuable site for exploring the relationship between human agency and ecological change. Like the matsutake mushroom forests analyzed by the anthropologist Anna Tsing, heathlands defy Western obsessions with pristine wilderness and its despoliation by supplying "models of well-being in which humans and nonhumans alike might thrive." They offer actual, real-life examples of sites where "something that might be called a sustainable relationship between humans and nonhumans could be imagined."[42]

In a sense, then, heathlands' hybrid nature only magnifies the loss incurred in their disappearance. The absence of English heathlands is a symptom of the damage modern life has inflicted on nature and culture alike—including the almost insurmountable difficulty of remedying such damages. Any attempt to restore English heathlands would require not only the remediation of environmental damage but also the re-creation of entire economies of common rights and local manufacturing driven to extinction more than a century ago. The loss of heathland represents the loss of one of the last sustainable interspecies communities widespread across Britain and the European continent.

If the special economic and ecological relations realized in English heaths have largely vanished, there are still ways of gathering some intellectual sustenance from this unappreciated landscape. The disappearance of these symbiotic communities was enabled by modes of storytelling that made it difficult to conceive of their disappearance as an actual loss. Literary treatments of English heathland from Shakespeare's *King Lear* (1606) to Emily Brontë's *Wuthering Heights* (1847) approach it as little more than an imposing backdrop, a conveniently stark and ominous setting for internal human drama. Despite his best efforts to reincorporate nonhuman nature into his social novels, Hardy continued to work within inherited distinctions between human character and nonhuman setting—distinctions that made the reality of ecological interdependence all but impossible to comprehend. The real-estate developers, foresters, and agronomists who pounced on heathlands inherited a different form of storytelling about communal dynamics: the doctrines of modern economics, which prize profit above all else in the exchanges that tie objects, animals, plants, and people together. Sustainable but unprofitable, the particular symbiotic process called heathland was economically interesting only insofar as it could be cheaply converted into something of higher cash value.

It took a naturalist to see what heathland actually was—but seeing it was not enough to save it. Darwin could discern the reality of heathland because he applied to it those strategies of ecological plotting so visible in the widely read works of Malthus and Martineau in the early 1800s. This use of ecological plots enabled Darwin to recognize in the bleak English heath a harmonious but fragile community, one co-produced by many interdependent species. His extraordinary storytelling about heathlands and other biotic communities created a coherent framework for the life sciences, opening up many of the avenues of research in biology and ecology that persist to the present.

Faced with such communities, however, Darwin had no ethical recommendation about what to do with them. He was a naturalist, not a moralist. And as Haeckel and other self-proclaimed Darwinians adopted Darwin's ideas and techniques, the life sciences only grew more rigorous, more professional, and more invested in objectivity—an objectivity defined by a refusal to associate science with specific political or ethical positions. What had happened to political economy in the early Victorian era happened to natural history by the end of the same century: it became an increasingly specialized, increasingly mathematical science, its once accessible and intriguing narratives replaced by mathematical postulates and technical terminology.[43] Before long even Darwin's works—the writings that almost single-handedly founded the fields of ecology and evolutionary biology—would look decidedly amateurish in this new context. They unselfconsciously weave together personal experience, parlor experiments, and collected anecdotes that charm and persuade without ever seeming quite scientific.

There were obvious advantages to the disciplinary rigor introduced to the life sciences after the Darwinian revolution. From the standpoint of intellectual history, however, they represented a decisive closing-down of the ethical possibilities latent in the cross-disciplinary, cross-species storytelling that showed such promise at the dawn of the nineteenth century—the storytelling I have been calling the ecological plot. Adopted and adapted in many fields, this essentially literary technology for imagining interconnection proved amazingly fruitful. With each modification of it, however, something was lost. The political economists who took up the ecological plot's vision of interconnection ended up focusing on commercial exchange alone. They relied on monetary value to settle questions of significance and priority in an interconnected world. The more ethically oriented storytellers who spun the ecological plot out into realist literary fiction aimed for a wider view of significance and priority. Yet their more expansive definitions of value played out within a more restricted definition of community, one that all but eliminated the nonhuman world from the imagination of interconnection. To their credit, the naturalists who focused on the dynamic relationships within the nonhuman world also began to map its entanglement with human society—but they generally refrained from assigning value, significance, or priority to such entanglements. They were committed to describing the interconnected world but did not think it proper to use their expertise to intervene in it.

As these three once-related fields sharpened their focus, they inevitably broke apart. The promise latent in the ecological plot—the promise of a new ethical imagination, one founded on the material interdependence of human and nonhuman communities—receded. In its place came the increasingly nuanced but discrepant visions of collective life offered by humane moralists, market analysts, and detached scientific observers. These views came to seem so distinct as to be totally unrelated, even irreconcilable. For a time, the benefits of specialization seemed to outweigh the conflicts and contradictions of these divergent approaches to knowledge and to ethics. Now, however, with communities worldwide facing the many "wicked problems" that cluster at the edges of traditional areas of social, economic, and environmental expertise, specialization increasingly appears as an obstacle in its own right. The need to somehow reconcile such approaches and work across disciplines is one we ignore at our own peril.[44]

CONCLUSION

Distant Relations

Charles Elton was only twenty-six years old when he wrote the book that would make his name and change his field forever. To be fair, Elton had a bit of a head start in the writing business. Literature was in his blood. His father, Oliver Elton, was a professor of English at Liverpool University whose many books, biographies, and translations included the monumental six-volume *Survey of English Literature, 1730–1880* (1912). His mother, Letitia MacColl Elton, had published several nonfiction works for children, including *The Story of Iceland* (1887) and *The Story of Sir Francis Drake* (1906). Evidently drawn to a life of letters, Charles would eventually marry the poet Edith Joy Scovell. Yet Elton himself had no pretensions to literary greatness. His groundbreaking book, *Animal Ecology* (1927), was intended only as an introductory textbook, a sort of preliminary synthesis describing the present state of a newly solidified science.[1]

After Haeckel coined the term *ecology* to describe Darwinian methods in 1866, interest in the field had grown steadily, particularly among botanists. The first textbook on plant ecology appeared in 1895, penned by the Danish botanist Eugenius Warming. Translated into English in 1909, it inspired a rapid growth in the study of plant communities on both sides of the Atlantic. The coming decades saw debates among British and American scientists about how plant communities changed over time, with particular interest in the progress of steplike changes known as succession—the sort of large-scale shifts in vegetation that Darwin had observed back on English heathlands in the 1840s and 1850s. But Elton's book, dashed off in just under three months, was effectively the first to integrate existing ideas into a single system, one that highlighted the messier, harder-to-observe contributions of animals to the dynamics of cross-species communities. This "capstone," as one historian puts it, "codified the organizing ideas" of the nascent field. "Elton's book *Animal Ecology*," another observes, "stands as a marker between the old natural history and the new."[2]

Natural history was, in fact, the banner under which Elton understood his own ecological work. Ecology was radical, he argued, only in its renovation of a long-established field. "Ecology," he explains on the first page of *Animal Ecology,* "is a new name for a very old subject. It simply means scientific natural history."[3] As Elton described it, the problem with natural history was that it had always been an amateurish, scattershot affair. Hobbyists wrote up personal anecdotes about animals and plants in widely dispersed venues, rarely taking care to accurately name and describe the species they observed. Meanwhile, professional zoologists and systematists spent their hours examining dead specimens, using comparative anatomy to classify their subjects and build out an admirably nuanced tree of life—but one that failed to account for the lived interactions among the species they named. Almost no effort had been made to bridge the gap between these two ways of practicing natural history. As a result, the scientific world lacked viable explanations of how plants, animals, and fungi came together into dynamic, functioning communities. Even into the twentieth century, literature offered more tantalizing clues to such relationships than science did. As Elton exclaims, "[T]here is more ecology in the Old Testament or the plays of Shakespeare than in most of the zoological textbooks ever published!" (7). Careful storytelling would be needed to bring the dry pedantry of taxonomy and the aimlessness of animal anecdotes together under the purview of a systematic science: "We have to face the fact that while ecological work is fascinating to do, it is unbearably dull to read about, and this must be because there are so many separate interesting facts and tiny problems in the lives of animals, but few ideas to link the facts together" (17).

To remedy the problem, Elton had to reunite modes of thinking that had drifted apart, parsed into different disciplines over the prior century. "When one starts to trace out the dependence of one animal upon another, one soon realises that it is necessary to study the whole community living in one habitat, since the interrelations of animals ramify so far," he observes (189). Following the details of other creatures' lives required recognizing not only the interdependence of individual organisms but also the interdependence of superficially isolated branches of knowledge. Before long, the student of ecology was forced to admit that "it might be worth while getting to know a little about geology or the movements of the moon or of a dog's tail, or the psychology of starlings, or any of those apparently specialised or remote subjects which are always turning out to be at the basis of ecological problems encountered in the field" (188).

From Elton's perspective, the most important interdisciplinary work necessary to ecology was the unification of economics and natural history. Economics and the social sciences had developed rigorous structural and mathematical approaches that promised to give order to the messy facts recorded by naturalists. If ecology was the new name for a more rigorous, systematic natural history of interspecies relations, he notes in the preface, ecology was essentially just "the sociology and economics of animals" (v). Economic analogies and metaphors figure regularly in Elton's depictions of interspecies communities. Thus each animal takes up its own occupation in the division of labor, a role that ties it to its fellows in exchanges of matter and energy. "Copepods," he remarks at one point, "are living winnowing fans, and they form what may be called a 'key-industry' in the sea" (57). At several points, the difference between capitalists and the lower classes helps explain the different ways members of the food chain gain their livelihoods. "Since it uses the bark of trees for food," Elton reports in one example, "the beaver is unaffected by annual variations of plant food-supply. It lives on capital and not on income—an almost ideal existence" (138). A discussion of stable population numbers of each species leads him to an analogy where animal communities function like a farm, as the operating capital (food) must be allocated properly among a particular number of laborers to maximize overall productivity (113).

Malthus, the luminary who had wed economics to interspecies relations in the late 1700s, is nowhere to be found in Elton's list of references. Nevertheless, Malthusian concepts underpin every aspect of his thought. "Food is the burning question in animal society," Elton reminds us, "and the whole structure and activities of the community are dependent upon questions of food supply. . . . There are, in fact, chains of animals linked together by food, and all dependent in the long run upon plants" (56). He echoes Malthus even more directly in his focus on populations and the different ways they experience "an effective check" to limit their numbers (111). As it turns out, Elton had absorbed Malthus the same way so many others had: secondhand. He took up Malthusian notions both through his readings of Darwin and even more directly through his rapid absorption of the writings of Alexander Carr-Saunders, a biologist who penned the neo-Malthusian book *The Population Problem* (1921) and eventually became director of the London School of Economics.[4]

Elton's synthesis turned his slim introduction to animal ecology into a cornerstone of the science. Julian Huxley, a former teacher of Elton's and the editor of the series where *Animal Ecology* appeared, praised the book as "the first in which the proper point of view of animal ecology has yet been explicitly stated."[5] Among his many contributions, Elton popularized the term "food chain" to describe the links of consumption that unite producers and consumers into an interconnected circuit. He charted the way energy dissipated over the course of food chains, producing a community structure where plants and other producers outnumber the herbivores that eat them, which in turn outnumber the predators that hunt them—a structure he called the "pyramid of numbers." Elton was also the first to define and elaborate on the idea of an ecological "niche," the term he used to describe an organism's role in the broader community.[6]

In all of Elton's careful tracing of interdependence and how to study it, however, there is something conspicuously missing. His stories of interconnection and even the metaphors he uses to describe them often recall the chains of relations narrated so painstakingly by authors of Victorian realist novels. "[T]he food-relations of animals are extremely complicated and form a very closely and intricately woven fabric—so elaborate that it is usually quite impossible to predict the precise effects of twitching one thread in the fabric," he writes, echoing the web metaphor favored by both George Eliot and Charles Darwin (79). Each interspecies community, Elton remarks, really deserves to be captured within the confines of a single, sprawling book: "One habitat alone, the edge of a pond, or the ears of mammals, would require a whole book if it were to be treated in an adequate way" (17). But never does he dwell seriously on questions of ethics. The term features only once in his study, in a final diagram that shows how ecology must necessarily draw on a variety of fields to derive a full picture of interspecies relationships. "Ethics" is there, clustered with "Medicine," beneath the question of how to understand parasites' role in the natural world (191). But for Elton, the ethics of parasitism were essentially settled. As members of the natural world simply trying to survive, parasites were no less moral than any other creature that derived its sustenance from the act of taking advantage of others. Any ethical objection to parasites would require a much larger condemnation of nature itself. In such a system "we must, to be consistent, accuse cows of petty larceny against grass, and cactuses of cruelty to the sun" (75).

Human effects on ecosystems were clear to Elton, but he did not see it as helpful or necessary to fret over their ethical dimensions. After noting the slaughter of bison, the extinction of the passenger pigeon, and the depredations of the whaling industry, Elton acknowledges an evident pattern. "Almost everywhere the same tale is told," he admits: "former vast numbers, now no longer existing owing to the greed of individual pirates or to the more excusable clash with the advance of agricultural settlement" (106). But Elton immediately abandons any claim to moral judgment. "It is not much use mourning the loss of these animals," he decrees. "Our object is rather to point out that the present numbers of the larger wild animals are mostly much smaller than they used to be" (106). Elton evidently admired the conservation movements that were emerging at the time. Within the confines of the book, however, they appear primarily as efforts whose consequences the ecologist should watch with an observant but dispassionate eye.[7]

The air of objectivity that surrounds this early ecological work is hard to divorce from its ongoing alliance with economics. That relationship was never purely intellectual. For ecologists, economics provided a model of a field that used mathematical laws to provide rigorous descriptions of community dynamics. But many ecologists also relied on economics to justify the very existence and value of their own field. With their growing knowledge of how plant and animal populations worked, ecologists promised to maximize productivity and minimize pest disruption in the many industries that relied on plant and animal resources. As the historian Donald Worster puts it, the early science of ecology was tied to a "managerial ethos" that treated the natural world as one more potential zone of exploitation to maximize wealth.[8] In Elton's case, the relationship was personal. The purely academic research he first undertook in the Arctic circle and the Outer Hebrides was soon supplemented by consulting work for the fur trade, followed by years devoted to applied questions of agricultural pest control.[9]

In short, early ecology leaned heavily on economics both as an intellectual model and as a source of practical applications that would justify investing in the otherwise complex, difficult, expensive research necessary to understand nonhuman communities. There were multiple powerful incentives for early ecologists to play nice with economics and the forces of industry. Denouncing the harvesting of natural resources in clear moral terms would have threatened the stability of the upstart science, the field that promised to reveal how to effectively manage and sustain such nonhuman communities in the first place. But some ecologists settled more happily into such economic partnerships than others.

Animal Ecology is peppered with hints about the implications of ecology for issues "of great economic importance" (141).[10] It is in Julian Huxley's introduction to the first edition, however, that the excitement over the applied economic prospects of ecology becomes most apparent. There, research in ecology is trumpeted as necessary "if man is to assert his predominance in those [tropical] regions of the globe whose climate gives such an initial advantage to his cold-blooded rivals, the plant pest and, most of all, the insect."[11] Ecology matters not purely as an intellectual endeavor but also—and most importantly, to the world outside the ivory tower—as part of this imperialistic civilizing mission. Only under the tutelage of ecology will "[t]he more advanced governments of the world" be able to ensure comprehensive mastery over "the control of wild life in the interest of man's food-supply and prosperity."[12] Later students of ecology would chafe under this insistence that scientists work hand in hand with the industrializing forces of economic development.

But ecologists did not turn out to be the first or only Western defenders of nature to take issue with the forces of industrialization. In the decades when ecology was still coming together as a scientific discipline, several grassroots movements had emerged independently to protest the exploitation of nature. Many early conservationists in Europe and America actually worked in forestry or land management, and their experiences taught them the importance of prudent restraint in the harvesting of natural resources. Another camp of nature defenders—the preservationists—took a more radical approach. Drawing on Romantic forms of nature veneration, they argued for the total protection of certain landscapes from all forms of use. For preservationists such as the Scottish-American writer and Sierra Club founder John Muir, treating wild places as reserves for water and timber was shortsighted and sacrilegious, like treating cathedrals as quarries for stone. Battles over the proper amount of veneration and restraint to show the natural world played out on both sides of the Atlantic in the late nineteenth century. In the 1870s, for example, the city of Manchester faced unexpectedly fierce opposition from people around the globe when its leaders began planning a reservoir in the Lake District, a part of England that William Wordsworth, Samuel Taylor Coleridge, and other Romantic poets made famous for its wild beauty. In the United States, the city of San Francisco sought for decades to dam the Hetch Hetchy valley in

Yosemite for the same purpose—and faced analogous outrage from admirers of the natural landscape. Ultimately, municipal water needs won out in both cases. But these fights showcased the growing sense that nature itself was a public good that should not be despoiled for the material needs of human beings.[13]

These revolts against urbanization and industrialization were undeniably ethical and nature-oriented. Nevertheless, they were not thoroughly grounded in ecological ideas of interdependence across species. Within a few decades of the publication of *Animal Ecology*, however, a new crop of scientists trained in the field began to recognize the importance of reintroducing ethical questions to the ecological networks that scientists were mapping with professional disinterest—and they began to join forces with adherents of these earlier movements. The most influential of these scientific advocates were American, but their work was informed by global conversations and had international import. Among these thinkers was Aldo Leopold, a game manager whose formative education included time abroad studying sustainable German forestry. Leopold was a generation older than Elton, but the two men struck up a friendship when they met at a Canadian conference on biological cycles in 1931.[14] Unlike Elton, Leopold had grave doubts about the consequences of the new, amoral ecology. He explored those doubts most fully in his posthumously published *A Sand County Almanac: And Sketches Here and There* (1949).

Looking back through the reasoning of conservationists in the first half of the twentieth century, Leopold lamented their continued subservience to economics. "One basic weakness in a conservation system based wholly on economic motives is that most members of the land community have no economic value," he explained. "When one of these non-economic categories is threatened, and if we happen to love it, we invent subterfuges to give it economic importance." The case of disappearing songbirds was the most striking example: "Ornithologists jumped to the rescue with some distinctly shaky evidence to the effect that insects would eat us up if birds failed to control them."[15] Even if such economic arguments sometimes held water, the problem was one of principle. As Leopold observed, for too long ecologists had been forced to cater to a restrictive definition of value borrowed from economists, despite ecology's growing coherence as a field. Ecology lacked its own robust formulations of value and significance, because in all practical policy applications, its representatives confronted a world in which "evidence had to be economic to be valid."[16]

The situation Leopold lays out should sound almost uncanny to anyone familiar with Victorian culture. Around the middle of the century, an emerging group of morally committed writers found themselves faced with a widespread crisis of valuation. The near-total dominance of economic thought was causing human beings to approach questions of interdependence in dangerously shortsighted ways. Those in power could only see value when it appeared in monetary form, and they were overlooking the kinds of reciprocal, ethical relationships key to both social stability and humane moral conduct. In essence, a writer like Leopold faced the same situation that had worried social-problem novelists a century earlier. Those Victorian writers had inherited a rudimentary notion of interdependence from the ecological plotting of Martineau and Malthus, but they had retreated from the almost impossibly complex ethical questions raised by ecological relationships. Leopold came from a later branch of the same Malthusian tradition. But by the time he wrote, the offshoot known as ecology was in full blossom. Leopold had robust conceptions of ecological succession, population dynamics, and predator-prey relationships at his disposal. They were complemented and honed by his years of experience working for the Forest Service and writing about game management—years that included a full reckoning with the emotional and ethical impacts of destroying nonhuman life.[17] So while he resembled the humane moralists of the past in his grave doubts about economic methods of valuing interdependence, he differed from them in one key respect: in such systems, he recognized the value of both human and nonhuman life.

In one of the first explicit statements of modern environmental ethics, Leopold urged all people to extend the idea of the moral community to our material entanglements with the nonhuman. Implicitly calling for a reunion of the three branches of Malthusian thought that had split apart, he identified the need to view interspecies networks of interdependence through three conjoined lenses. In approaching other species, the ethic and the aesthetic had to be reunited with the economic to produce a truly environmental approach. "Examine each question in terms of what is ethically and aesthetically right, as well as what is economically expedient," he wrote. "A thing is right when it tends to preserve the integrity, stability, and beauty of the biotic community. It is wrong when it tends otherwise."[18]

By the time Leopold's trailblazing book appeared, others were already bushwhacking their way toward the same path. The chief editor of publications at the U.S. Fish and Wildlife Service, for example, was busy building a moral case against the global pest management industry. Her name was Rachel Carson. During her graduate study at Johns Hopkins University, Carson had come across a slim volume on animal populations that convinced her that nonhuman life should be understood as its own sort of a community. The book was Elton's *Animal Ecology*. It was not until the end of World War II, however, that Carson turned her attention fully to the dangers human activity might pose to such communities. After more than a decade of writing and researching, she published *Silent Spring* (1962), the book that is widely credited with launching the modern environmental movement.[19]

In many ways Carson's writing departs from Leopold's. What they share, however, is more important. Both express a deep-seated discomfort with the intimacy of ecological, agricultural, and economic interests. Leopold singled out the total dominance of economic logic as the biggest obstacle to full ecological understanding. Carson's book was far more focused and concrete in its resistance to industry. In it, she traces how the manufacturers enlisted in the war effort during the 1940s shifted to different modes of killing in the postwar era, developing an ever-expanding range of chemicals to destroy agricultural pests. Yet these chemicals, she explains, were promoted among businesses and consumers without serious investigation into their long-term effects on other creatures—including the larger, hardier human beings routinely exposed to them—despite the fact that the substances were known to accumulate in the tissues of the body.

While they are remembered today for their revolutionary effects on environmental thought, Leopold and Carson were both realists in the political sense of the word. They took a more or less pragmatic approach to ecological concerns. Neither dreamt that economic interests would be permanently sidelined by more immaterial ethical and aesthetic ones. Carson, for example, is typically hailed as the advocate whose work enraged chemical manufacturers and led to a near-total ban of the pesticide DDT in much of the developed world. It is true that many in the chemical industry were furious about the publication of *Silent Spring*; they repeatedly tried, and repeatedly failed, to smear the naturalist's public reputation. Yet Carson herself never called for outlawing DDT, and the ban popularly attributed to her work would not be enacted until a decade after her book's publication.[20] She strikes a more moderate tone throughout *Silent Spring*. There she objects to the careless blanket spraying of pesticides and herbicides while still accepting and even advocating "a perfectly sound method of *selective* spraying," which she allows has "many advantages."[21]

Leopold is similarly conciliatory toward his opposition. Immediately after laying out his dream of an environmental ethic, Leopold acknowledges the reality that such an ethic will always be constrained by some economic dictates: "It of course goes without saying that economic feasibility limits the tether of what can or cannot be done for land. It always has and it always will."[22] Even within such constraints, however, he insists that the widespread cultivation of a land ethic would have a transformative effect on human behavior and the ecosystem as a whole. It all comes down to rejiggering the popular imagination to make room for a new ethical vision of community. "An ethic to supplement and guide the economic relation to land presupposes the existence of some mental image of land as a biotic mechanism," Leopold insists. "We can be ethical only in relation to something we can see, feel, understand, love, or otherwise have faith in."[23]

These moderate appeals to conscience, paired as they are with calls for a renewed imagination of community, suggest another sense in which the new crop of environmental writers appearing in the mid-twentieth century might be called realist. They resemble their Victorian realist forebears in their practical, commonsensical pleas for basic human decency—both as an intrinsic good and as a step toward communal self-preservation. What's more, they drive home those appeals through the strategic use of imaginative writing, drawing on literary and fictional methods to help readers understand the very real material connections that bind communities together.

Leopold's *Sand County Almanac* is remarkable for the way it gropes toward a genre or literary form adequate to capture this message of interconnection. Part 1 of his book adopts the popular conventions of the nature journal—the almanac of the title—detailing monthly changes on an abandoned farm that Leopold was restoring from the brink of ecological exhaustion. Part 2 roams more widely, featuring essays about different landscapes around North America that sometimes veer abruptly into more fictional and conceptual terrain. *A Sand County Almanac* makes a final decisive turn away from the descriptive sort of nature writing in part 3, where Leopold tries to distill the philosophical lessons of his writing and lay out how a land ethic might work.

Where literary form is concerned, then, the book is a mess. But that mess matters. It shows an ecologically and ethically attuned writer fumbling for the right narrative shape to transform popular intuitions about nature and society. He is struggling toward a form that will capture his basic understanding of "land as a community to which we belong."[24] So, in a section he simply calls "Odyssey," Leopold lays out the stories of two characters named X and Y as they move through an ecological community bound by transfers of matter and energy. Yet X and Y are not characters in any traditional sense of the term. They are atoms, experiencing nothing as they are unconsciously shunted across the networks of interdependence over time. X, for example, moves through an idealized premodern community from his prehistoric resting place in a "limestone ledge" to a "bur-oak root" to "a flower, which became an acorn, which fattened a deer, which fed an Indian, all in a single year."[25] Y is routed through an industrialized agricultural complex that leaves him literally adrift after profit-driven forms of agriculture and engineering dump him uselessly into an ocean from which he cannot be recovered. Although Leopold doesn't seem to know it, "Odyssey" revives an eighteenth-century tradition of storytelling about personified objects—a long-dead genre of tales that scholars call *it-narratives*—and tries to retrofit the form with an element of ecological education.[26]

Carson's work shows evidence of similar struggles with genre. *Silent Spring* opens with "A Fable for Tomorrow," the story of a happy, harmonious rural town suddenly cursed with widespread death and suffering. Birds vanish from the trees and fields, livestock die off, well-stocked rivers suddenly grow empty, humans sicken, pregnancies cease. This "evil spell" is accompanied by the mysterious descent of "a white granular powder" that falls across the landscape.[27] "I know of no community that has experienced all the misfortunes I describe," Carson confesses. "Yet every one of these disasters has actually happened somewhere, and many real communities have already suffered a substantial number of them. . . . [T]his imagined tragedy may easily become a stark reality we all shall know."[28] The rest of the book works backward, like a sort of time-traveling murder mystery, as Carson races to link these scattered disasters back to their sources and so prevent the horrific fairy-tale future that opens her story. She shows how a postwar industrial complex rapidly produced new chemicals and brought them to market to kill insect and plant pests, never considering the ecological flows of matter and energy between species who share overlapping environments and participate in the same food webs. Following the consequences of these chemicals takes her from wartime laboratories to the fields and forests where spraying takes place, through the chains of interlinked creatures affected, and finally into the recesses of the human body, where recent advances in microbiology were just beginning to shed light on the origins of cancer cells and the power of mutagenic chemicals to create them.

To anyone schooled in Malthus, or Darwin, or simply in the storylines of the great Victorian multiplot novels, *Silent Spring* hums with unexpected resonance. Take for example Carson's story about John Mehner, a graduate student at Michigan State University who decided to do his doctoral thesis on populations of the American robin (*Turdus migratorius*) on the school's campus—in part, presumably, because the bird was so ubiquitous and close at hand. The study began well enough in 1954 with a robust robin population. In 1955 the robins returned for another nesting season. That year, however, the corpses of fallen robins began to proliferate on the quads. Soon nests and young dwindled. The research of Mehner, his professor George Wallace, and the scientist Roy Barker (engaged in similar work at another institution) eventually revealed what was going on: the robins were dying of DDT poisoning. But the actual explanation of how the robins were poisoned is far more complex and extended than it appears. Like Darwin's realization that cattle create heathland, or Malthus's story of horse fanciers causing starvation among impoverished laborers, or Gaskell's tale of a cold winter leading to political assassination and factory reform, understanding the robin die-off was not straightforward. It required a much deeper investigation into the chains of cause and effect that emerge from overlooked forms of material entanglement. The story of the Michigan State robins began decades earlier, with a stowaway in the cargo of the international shipping industry.

Sometime in the late 1920s or early 1930s, a load of burl logs arrived at the Port of New York, destined to be reworked into veneers. A burl is the technical name for the curled, wavy, uniquely patterned sections of wood that form at the base of some trees as an adaptive response to stresses like fire or fungal infection. The beauty and idiosyncrasy of burlwood makes it a natural choice for the exterior panels of walls, doors, and other furniture. In this case, though, the burl logs carried their fungal infection with them. After they arrived, boring insects carried the fungal spores from the dead logs to living American trees. The sawing of such trees into timber and firewood hastened the fungal spread. The result was Dutch elm disease, an epidemic that expanded over the next several decades to wipe out virtually every mature native elm in the United States. In the midst of this epidemic, local and state authorities across the American Midwest tried to contain the disease and save the elms by the blanket spraying of DDT, a pesticide they hoped would kill the bark beetles responsible for the disease's transmission.

Spraying on the Michigan State University campus began in 1954, the first year of Mehner's population study.

In a straightforward poisoning plot, the fate of the robins that year would have been sealed. But this was no straightforward plot. The DDT did not kill the robins that year. Instead, it coated the leaves of the trees and remained there—harmless, apparently, to the birds that flitted among them. With the arrival of cold weather, the robins departed in roaming flocks as they did every year. The leaves fell as usual and blanketed the earth. The natural world seemed unaffected, idyllic. When the weather warmed again the next spring, ecological cycles began anew. The campus's earthworms emerged from the slime-coated balls underground where they lay dormant. Revived from their seasonal torpor, they began their regular work of pulling leaves into their burrows and eating them. This year, however, something was different: the fallen leaves were still coated with DDT from the last year's spraying. The worms ingested the insecticide. The robins returned and ingested the worms. Eventually, after the birds had eaten enough worms, the accumulated effect of the DDT in their bodies sterilized or killed them a year after the spraying.[29]

Carson's meticulous tracing of chemical poisons as they traveled through the environment essentially reintroduced ecological plotting to the popular imagination. It shed new light on what she—recalling her Victorian predecessors—described as "a vast web of life, all of which needs to be taken into account."[30] The reality of material exchange and the interdependence it produces had been sunk under mathematical equations and anthropocentric storytelling for a century. Now it was once again an urgent ethical concern, because the matter being exchanged was explicitly designed to kill the animals that came in contact with it.

Previous stories of the exchange of matter and energy had required some imaginative heavy lifting, because most plants and animals radically transform the matter and energy they absorb from others. The fact that sunlight, water, and carbon dioxide combine to become sugar in the leaves of a plant is impossible to witness firsthand; it requires belief in the invisible molecules of the air having a certain structure and combining almost alchemically with both the water taken into the plant and the energy absorbed from the sun. Even the way that plant material becomes animal flesh, and the way that eating literally transforms one being into another, has a certain wizardry to it. Similarly, the processes through which the exchange and transformation of matter creates money, and the ways that money can be transformed back into various goods and services, all possess a tinge of transubstantiation. The entire history of economics is basically a series of mathematical arguments about how, exactly, such baffling transformations work.

Chemicals like DDT were different. Their horror lay in their failure to transform over the course of such exchanges. While DDT does eventually degrade into other chemicals, it takes years—and the right environmental conditions—to do so. In most cases, as DDT moved from factories to tree leaves to worms to robins, it never altered its form or function. It remained DDT, a molecule capable of causing tissue damage or death to any animal that ingested it. The nonfungibility of manufactured pesticides kept them dangerous as they moved between species, creating a "chain of devastation" that dramatized the ties that bind animals, plants, soils, waters, and human beings together.[31] The chain's tragic aspect thus came with an unexpected upside. It was no longer necessary to tell fictional stories about molecules named X and Y to imagine where and how matter traveled. Precisely because these substances regularly killed the organisms that materially incorporated them, it became comparatively easy to track every place—and every being—they had been. And whenever they were suspected as a cause of such die-offs, the chemicals' unaltered structure made it possible to test for their presence and concentration in nonhuman bodies and inorganic surroundings.

With the advent of persistent organic pollutants, then, the imaginative leaps necessary to envision interdependence were shortened to a set of comparatively simple and uncontroversial steps. The very robustness of these new man-made poisons had made ecological interdependence both visible and morally urgent. Even the most self-interested reader could not fail to be unsettled by the ecological plots sketched out in *Silent Spring*, because they concerned not just the suffering of birds and fish but also the suffering and death of human beings. These human beings could be anyone—they could be you and me, and we might not even know it. The victims of persistent organic pollutants had little knowledge of their own exposures, little control over whether they were exposed, and little certainty about how such exposures would damage or destroy their lives. What Carson's stories showed was that the chain of devastation forged by chemical pesticides may originate in human society and be applied to the nonhuman natural world, but its effects do not stop at some imaginary line dividing nature from society. The same poisons that kill off a few generations of insects and devastate bird and fish populations also settle into the tissues of human beings. There, they gradually accumulate until they reach concentrations high enough to cause death or disability in unwitting victims decades later—victims who rarely connect the dots between the chemical exposures of their distant past and the ailments that surface many years afterward.

Carson may have been among those victims. In the midst of writing the chapters of *Silent Spring* detailing the ways chemical pesticides could cause cancer, she received her own cancer diagnosis. In a final, ironic coda to the book, the environmental luminary who composed it would die of complications from the very disease she warned about.[32] *Silent Spring* showed the public that knowing how to tell ecological stories, how to read them, and how to think using them was neither purely an intellectual issue nor purely an artistic one. It was a survival skill. The ecological plot is the story of our lives, whether we like it or not. Recognizing that you are a small but vital character in a much larger ecological storyline is literally a matter of life and death.

Meanwhile, where was literature? If environmentalists were rediscovering the value of ecological plotlines and the ethics of interdependence, were novelists and fiction writers also rediscovering the importance of cross-species relationships to a full understanding of community? For the most part they were not. The storylines they were invested in had left environmental factors behind long ago.

By the twentieth century, novelists had largely abandoned big-picture attempts to map cross-species connections and their ethical significance. This book has already traced how the early nineteenth century's open-ended understandings of interconnection became the ethical concern of fiction writers. It has shown, too, how authors briefly entertained such notions before pivoting to human society as the primary moral concern of realistic fiction. If anything, this narrowing of literary focus only sharpened after the turn of the century. The panoramic, almost sociological depictions of community widespread in Victorian fiction mostly died out. They gave way to nuanced investigations of the twists and turns of individual consciousness and even of the practice of novel-writing, hallmarks particularly prevalent in a new and newly self-conscious artistic movement called modernism.[33]

The modernists' disregard for realist methods and their concentration on individual minds did have some positive consequences for humans' attention to other creatures. As the critic Caroline Hovanec has shown, modernist writers remained in close contact with twentieth-century animal ethologists who were weighing what humans could and could not rightly say about animal minds. These interdisciplinary encounters had transformative impacts on how writers understood what minds were, what forms of sentience might be shared across the animal kingdom, and what human thinkers owed to their nonhuman counterparts. Hovanec concludes a recent study of this interdisciplinary traffic by noting that modernist strategies for understanding animal minds live on in twenty-first-century science writing—with at least one significant shift. Current treatments of animals include not just consideration of their complex psychological otherness but also the ecological outlook that modernist writing missed, particularly "a growing recognition of [animals'] fragility" in the face of extinction.[34]

The modernists' failure to ruminate on questions of extinction is hardly surprising, given that their explorations of animal subjects were not, finally, part of any larger moral investigations into questions of interdependence. Modernists were not seriously interested in animals and plants as members of a materially interconnected ethical community. Recent attempts to unearth an ecological modernism are much like recent attempts to find ecological insights across Victorian literature: through impressive, counterintuitive feats of interpretation, critics can and do locate previously unnoticed environmental implications in these works of art. There is nothing wrong with such ecological rereadings of the past. They draw valuable attention to the subtle ways environmental issues can register within the cultural tradition. In the process, they add to our understanding of environmental history and our appreciation of the complexity of literary art.

There is a difference, however, between works that somehow include the nonhuman world and works that demonstrate a substantive understanding of natural interactions that qualifies as ecological. The former are common; the latter, rare. Confusing the critical practice of ecological reading with the historical practice of identifying milestones in the development of ecological ideas creates a problem. It produces a misleading picture of the cultural past at the very moment when a clearer sense of ecology's history is necessary to help us understand the modes of thinking that led to the environmental crises of the present. If anything, the belated scholarly effort to sift through Victorian and modernist texts with painstaking care to discover instances of environmental engagement is an indication of the marginal status such works assign to the natural world.

There is no need to condemn such works for their omissions. But it is risky to retroactively assign them ecological and political significances they did not have; an honest historical reckoning with their oversights is more valuable. A clear definition of what constitutes recognizably ecological thought—one grounded in cross-disciplinary reading in ecology and environmental history, as distinct from looser notions of nature appreciation, nature worship, and environmental feeling—helps remedy this confusion between ecocritical reading and ecological writing. This book has proposed one such definition. It identifies ecological thought as a particular recognition of material interdependence across species that arises from a distinct form of narrative storytelling. This definition emerges from extended engagement with literary works and with milestones in ecological thought. By these standards, the mounting scholarship devoted to ecocritical reading still does not reveal anything like a comprehensive, self-conscious ecological ethic running through the modern Anglophone literary tradition—for the painfully simple reason that no such ethic is there.[35]

Modernism was not the only literary development in the first half of the twentieth century, of course. With their in-depth investigations of the disorderly human consciousness and the even more esoteric matters of literary form, modernist works like Djuna Barnes's *Nightwood* (1936) and James Joyce's *Ulysses* (1922) were artistically respectable but, for many lay readers, all but incomprehensible. The same era saw the rise of much more popular genre fiction, a sprawling category that encompasses horror, fantasy, the detective novel, the western, weird fiction, and others. None of these modes foregrounded realistic, morally earnest depictions of interdependent community, though weird fiction did linger (as Hardy had) over the uncanny power and presence of nonhuman beings hidden somewhere in the landscape.[36] At times detective fiction dabbled in matters of the nonhuman, too. Capitalizing on the literary tendency to overlook other creatures, writers of detective stories sometimes employed the animals hidden in plain sight as crucial evidence in criminal plots, a tactic on display in works like Arthur Conan Doyle's "Silver Blaze" (1892).[37]

The ascension of modernism and of genre fiction never entirely unseated realism as a dominant literary mode, however, and the realism bequeathed to the twentieth and twenty-first centuries was almost entirely confined to the examination of human relationships. Those fringe works that achieved popular success in their committed portrayals of nonhuman life—including books like Ernest Thompson Seton's *Wild Animals I Have Known* (1898), Felix Salten's *Bambi* (1923), Henry Williamson's *Tarka the Otter* (1927), and Fred Bodsworth's *Last of the Curlews* (1954)—certainly contributed to growing concern over the plight of nonhuman animals in the modern world. Yet they never possessed the panoramic ambition to map entire communities so common in their Victorian predecessors, and they remained marginal backwaters to mainstream literary production. When these animal tales could not be justified through allegorical parallels of human experience, they risked dismissal as unrealistic, childish books that excessively humanized animal life. Literature remained what it always has been, a heterogeneous mixture of different types and styles of writing. Still, the detailed narrative treatment of plants and animals tended to serve as a line in the sand, an informal way of dividing fantastic imaginings intended for children from supposedly more elevated adult fiction.[38]

Yet the nonhuman was never completely eradicated from serious storytelling. Its place now is best captured by the haunting sound of a distant dog. In a comic but insightful 2010 essay, the writer Rosecrans Baldwin pointed out the shockingly pervasive presence of such dogs in realist fiction. Their survival is propagated through literature in a single sentence, slightly altered, that crops up repeatedly across the fictional canon. "Novelists," Baldwin observes, "can't resist including a dog barking in the distance." This distant dog never actually appears in the storyline. Instead, the animal surfaces—not unlike the metaphorical animals in *Middlemarch*—as a marker of distance itself, a kind of reminder of an otherwise excluded reality that the novelist is vaguely aware of but will never directly address. Writers seem at once compelled to mention this imaginary dog and oddly unaware of their participation in a long tradition of doing so. "Having heard the dog's call" as a reader, Baldwin says with astonishment, "it seemed like I couldn't find a book without one." Finding the invisible dog's voice in novels ranging from Leo Tolstoy's *The Death of Ivan Ilyich* (1886) to bestsellers by Jodi Picoult, Baldwin becomes increasingly paranoid, convinced that the barking dog is some kind of novelistic meme, an inside joke he does not get. He discusses the situation with a friend, another writer, who laughs at the idea—until the friend looks back over his own debut novel to discover that he has also included the untraceable bark. The pattern is complete when Baldwin returns to his own draft to find the distant dog there, too, unintentionally incorporated in his own prose. "These howls are empty and cheap," Baldwin concludes, "and I'll float the opinion that publishers should collar them."[39]

The intellectual genealogy traced in this book suggests otherwise. This unexpected convention is more than a pointless, amusing accident. Each faraway dog serves as a vestigial reminder of the social novel's ecological origins, a throwback to a not-so-far-off time when literary treatments of the significance of human lives crossed paths with, learned from, and absorbed related modes of writing about dependence between species. These modes gave rise not just to the realist novel but also to the modern disciplines of economics and ecology—the most rigorous tools we have for navigating ongoing questions of how humans relate to the wider world.

I hope this entangled history is interesting in its own right. It highlights the fact that the apparently discordant approaches modern humans use to explain our place in the world share a recent common ancestor. But recognizing this shared inheritance, and appreciating how such approaches broke apart, also suggests a new starting point for cross-disciplinary conversation. A nuanced understanding of how and why our dominant cultural, economic, and scientific modes of mapping community split apart clarifies their respective strengths. It also draws attention to their manifold shortcomings. It reminds us that the internal logic of a great novel, or an economic formula, or a scientific explanation feels convincing, coherent, and final only insofar as it tacitly reduces the complexity of the world, lulling us into believing we have a firm grasp on the messiness of our interconnected lives.

This study has offered a brief sketch of how different forms of writing have made such interconnected lives visible in the modern era. It has shown how literary techniques—in particular that strategy of assigning agency and significance that we call character, and the complementary strategy of aggregating things into stasis and insignificance we call setting—at once shape and constrain our ability to comprehend such interconnection. These modes of including and excluding function at some level across all disciplines, not just literature; the fields of ecology and economics are cases in point. But the social and natural sciences have a tendency to downplay the foundational role of language in their modes of thought. Over the years, the narratives and metaphors that underlie both economics and ecology have been obscured by a rapidly expanding superstructure of graphs, computational models, and mathematical formulae. These modes of representation increase certain kinds of precision, but they do so at the expense of accessibility, of big-picture thinking, and of a certain degree of humility, too.

When we fail to see the stories we have learned to tell ourselves for what they are—incomplete and imperfect stories, for all their value—we are unlikely to examine them critically to see what they may be overlooking. As a result, the frameworks employed by expert economists, ecologists, and members of the public often clash and produce unintended consequences. The stories of ecology, for example, are dominated by narratives of crisis, catastrophe, and apocalypse. At their best, they have the power to motivate people to see environmental problems they would otherwise overlook, teaching people how to care deeply for the nonhuman world. But their unrelenting emphasis on ecological loss and global disaster can also lead to depression, compassion fatigue, and hopelessness, as the scale of ecological problems seems too great for any individual to contribute toward a meaningful solution. These disaster stories also focus on human-generated problems that vilify contemporary society and those plants and animals many people know and love best—cats and dogs, farm animals, plants used in gardens and landscaping, and other introduced species. It is a strategy that risks alienating many people and even turning environmentalists themselves into misanthropes who cannot imagine positive, uplifting ways modern humans might relate to the natural world. Much of the focus of modern environmentalism involves a combination of urging people to stop doing things they enjoy and fighting for larger entities like governments to intervene in human behaviors.[40]

Given the entangled histories of ecology and economics, it may not be surprising that the stories of economics also veer toward crisis and apocalypse—but with a very different cast of heroes and villains. For many economists and proponents of business interests, the same forms of personal restraint and government intervention urged by environmentalists are understood as disasters in their own right, economy killers that threaten to bring about catastrophic recessions, job losses, and unpredictable shifts in global politics. Monetary prices remain the only consistent way of measuring value and significance within economic discourse, and the prospect of endless future growth—a growth that has no qualms about converting every last shred of the nonhuman world into economic goods and resources—appears like the only surefire way to outrun the persistent specter of economic collapse. As it is commonly taught, this neoclassical economics continues to see human beings (artificially divided into the simplified categories of producers and consumers) as the only significant agents or characters populating its world; environmental concerns are marginalized into the setting, demoted to externalities that lie outside the market system that economists study. The results of this reductive economic modeling are then touted politically as matters of fact, as inescapable economic realities.[41] And so, spellbound by discrepant worldviews, we struggle to reach satisfactory solutions in disputes over climate change, game hunting, resource rights, habitat destruction, and other wicked problems of the present that lie in the borderlands between social and environmental issues.

Reuniting these modes of thinking will be crucial to addressing the seemingly intractable problems of the present. Hopeful signs of progress are already springing up in the interstices where these fields meet—standpoints from which each isolated field can be questioned, tweaked, and refined. Ecological economists are interrogating the supposed necessity of endless growth for human flourishing, reigniting debates about sustainability and steady-state economies once associated with Victorian thinkers like John Stuart Mill. Behavioral economists working at the intersection of economics and psychology are eroding the classical notion of *Homo economicus*—the stereotyped main character of economic discourse, the consuming laborer who always makes choices with supremely self-interested rationality.[42]

This book suggests that the humanistic study of history and language also has a key role to play as we work toward a just and sustainable future. In the face of interdisciplinary gridlock, it helps to return to language as a literal lingua franca. Recognizing the linguistic and even literary base for our divided branches of knowledge—the *logos*, or word, at the heart of every *-ology*—emphasizes common ground and opens up new ways of talking across such boundaries. It reminds us that seemingly incontestable economic and ecological realities are themselves the products of discipline-specific stories and models, powerful but limited conventions that decide who and what is acknowledged, included, and prioritized. Literature and culture provide nonexperts with their own internalized stories, working theories of the world that shape our everyday sense of what morality and justice look like. A careful examination of our most essential stories—whether they emerge from our disciplinary training or from our larger cultural context—can remind us what they do and do not see. It offers one route toward the kind of *"cognitive opening"* that the theorist Sylvia Wynter argues will be necessary if members of different fields are to cultivate the necessary art of "questioning . . . the *self-evident unchallengeable unassailable truths* of their genre-specific storytelling."[43] At the same time, it promises to make us all more understanding of the stories that have proven essential to the worldviews of those who disagree with us.

This book has sketched out the genealogy of one such story. It has shown how that story's adoption and modification by different thinkers gave rise to conflicting notions of interconnection and how it ought to shape communal behavior. A more self-conscious approach to the stories we tell in every field—one that pays tribute to their shared histories, their unique strengths, and their respective limitations—offers a useful tool for easing political conflict, forging new coalitions, and enriching our collective picture of how the world works. It moves us past what the political scientist Claire Jean Kim has called *"single-optic vision,* a way of seeing that foregrounds a particular form of injustice while backgrounding others," and takes us toward "a practice of *multi-optic vision,* a way of seeing that takes disparate justice claims seriously without privileging any one presumptively."[44]

The social and natural sciences aren't the only fields that could benefit from a more thoughtful examination of how their stories work. Literature could, too.[45] Another word for telling stories is *relating,* and for good reason. Telling the right sort of story can help us recognize our intimacy with those whose perspectives seem distant from our own. Stories can help us bridge apparently impassable gaps, whether they separate the perspectives of different fields or the perspectives of different species. It may be that the trope of the distant dog really is a problem for modern literature, but not because the dog is an empty cliché. It is not the dog that needs to be reconsidered. It is the distance.

Notes

INTRODUCTION

1. Despite commonsensical distinctions between the arts and the sciences, the assumption of close intellectual traffic between literature and science underlies much of the literary and historical scholarship performed today. For a landmark collection that solidified the significance of this kind of interdisciplinary analysis for nineteenth-century studies in particular, see G. Levine, *One Culture*. There is a huge body of scholarship on the rise of the modern disciplines. For a history of changes in the arts and sciences in the early nineteenth century, see Klancher, *Transfiguring the Arts and Sciences*. For more on eighteenth- and nineteenth-century British practitioners of natural history, see Allen, *Naturalist in Britain*. For more on the poetic output of Victorian practitioners of science, see Brown, *Poetry of Victorian Scientists*. For useful overviews of the rise of science as a profession, see Barton, "'Men of Science'"; Daston, "Academies"; Sydney Ross, "*Scientist*"; and Yeo, *Defining Science*, chap. 1. For how objectivity came to define scientific standards and distinguish science from the arts, see Daston and Galison, *Objectivity*. On the rise of quantitative, scientific reasoning in social fields, see Porter, *Trust in Numbers*. For an analysis of how pragmatist pedagogy became the base of a (supposedly) unified scientific method, see Cowles, *Scientific Method*.

2. The son of a carpenter was William Whewell; see Yeo, *Defining Science*, 15. The literary critic was George Henry Lewes. For his legacy in physiology, see Haight, *George Eliot*, 255n5; and Smith, "George Henry Lewes." The role of novels in explaining social and ecological phenomena will be explored throughout this book.

3. Caroline Levine's arguments for the power of literary form to afford certain kinds of thought and action were a formative influence on this book; for her fullest elaboration of them, see C. Levine, *Forms*. But she is not the only literary scholar to theorize affordance; affordance theory began to enter literary study from multiple directions in 2008–9. The first introduction appears to come from C. Namwali Serpell, who draws on the concept as deployed in James J. Gibson's work. See Serpell, "Mutual Exclusion." In the next two years, Levine, Julka Almquist, and Julia Lupton joined Serpell in introducing the concept to literary studies (all seemingly independently). See C. Levine, "Narrative Networks"; Almquist and Lupton, "Affording Meaning." For a more detailed explanation of hypocognition, see Wu and Dunning,

"Hypocognition." There are already a number of excellent rhetorical studies that explore the relationship between language and ideas in the fields touched on in this book, though none of them trace the historical dissemination of a single literary form in the way I attempt here. For an accessible introduction to some of the ways in which language shapes thought, see Lakoff and Johnson, *Metaphors We Live By*. For discussion of some of the rhetorical and literary features of modern economics, see McCloskey, *Rhetoric of Economics*. For an overview of the rhetoric of biology and an investigation of some twentieth-century controversies, see Myers, *Writing Biology*. For a historical look at the genre and publishing format of the scientific journal, see Csiszar, *Scientific Journal*.

4. Commoner, *Closing Circle*, 33. Following in Commoner's footsteps, the theorist and literary critic Timothy Morton has argued that the idea that "everything is interconnected" is the central insight of ecology, dubbing this idea "the ecological thought." Succinctly, the argument of this book is that the ecological thought was preceded and made thinkable by the ecological plot. See Morton, *Ecological Thought*, 1.

5. Leopold, *Sand County Almanac*, 252–53. Leopold influenced both popular environmentalism and the entrance of environmental ethics into philosophy. For more on Leopold's life and influence, see Flader, *Thinking Like a Mountain*, esp. chap. 4; and Dunlap, *Saving America's Wildlife*, 65–110. For a landmark paper introducing Leopold to academic philosophy, see Callicott, "Animal Liberation."

6. For a classic and accessible account of story and plot in these terms, see Forster, *Aspects of the Novel*, 86.

7. Commoner, *Closing Circle*, 21.

8. As ecocriticism has emerged as a major field of research in Victorian studies, canny critics have begun to do exactly this with Brontë. See, for example, Thomas H. Ford, "Punctuating History Circa 1800," in Menely and Taylor, *Anthropocene Reading*, 78–95; Fuller, "Seeking Wild Eyre"; Gold, *Energy, Ecocriticism*, chap. 6; Kennedy, "Breath of Fresh Air"; R. Miller, "'Resolute, Wild, and Free'"; Pizzo, "Atmospheric Exceptionalism"; and Shawna Ross, *Charlotte Brontë*.

9. The breadth and depth of available ecocriticism is now too extensive to cite. For useful overviews, see Buell, *Future of Environmental Criticism*; Clark, *Cambridge Introduction*; and Garrard, *Ecocriticism*. For influential ecocritical studies of Romantic poetry, see Bate, *Romantic Ecology*; and Bate, *Song of the Earth*. For an important critique of that line of Romanticist ecocriticism, see Morton, *Ecology Without Nature*. For a classic ecocritical study of American transcendentalism, see Buell, *Environmental Imagination*. For energy transitions in early modern poetry, see the brilliant critical interventions in Menely, *Climate*. For a sampling of important ecocritical studies by Victorianists focusing on theories of energy, on the London fog, and on resource extraction, see, respectively, MacDuffie, *Victorian Literature*; Gold, *Energy, Ecocriticism*; Taylor, *Sky of Our Manufacture*; and E. Miller, *Extraction Ecologies*.

10. For an older but still useful overview of disputes surrounding the origins of ecology, see McIntosh, *Background of Ecology*, chap. 1. For a groundbreaking study of the precursors, including writers, to environmental ideas in natural history and natural philosophy, see Worster, *Nature's Economy*. For the importance of colonialism

and empire to proto-ecological concerns about resource scarcity, see Grove, *Green Imperialism*. For overviews of industrialism and the rise of conservationist thought within Britain, see Ritvo, *Dawn of Green;* Winter, *Secure from Rash Assault*. For the comparatively belated rise of the idea of the ecosystem itself within the natural sciences, see Golley, *History of the Ecosystem Concept*.

11. For the famous first-person account of Malthus's political economy crystallizing Darwinian thought, see Darwin, *Autobiography*, 98–99. For Marx both admiring and scoffing at Darwin's political economic influences, see McLellan, *Karl Marx*, 565. For a useful survey of claims about political economy and Darwin that ultimately rejects the importance of economic thought, see Gordon, "Darwin and Political Economy." For repeated attempts to build intellectual fields on the synthesis of Darwin and economic ideas, see (among countless others) Wilson, *Sociobiology;* Galor and Moav, "Natural Selection"; Hodgson and Knudsen, *Darwin's Conjecture;* Nelson and Winter, *Evolutionary Theory of Economic Change;* and Vermeij, *Nature*. For conflicts between political economy and nineteenth-century writers—especially realist novelists—see, most influentially, R. Williams, *Culture and Society;* Bodenheimer, *Politics of Story;* Cazamian, *Social Novel in England;* and Poovey, *Making a Social Body*. For the traffic between natural history and nineteenth-century literature, see, for example, Beer, *Darwin's Plots;* Duncan, *Human Forms;* Griffiths, *Age of Analogy;* King, *Divine in the Commonplace;* G. Levine, *Realism, Ethics and Secularism;* and Shuttleworth, *George Eliot*.

12. R. Williams, *Country and the City*, 165; C. Levine, *Forms*, 122; Beer, *Arguing with the Past*, 13, 15; Choi, *Anonymous Connections*, 57; Garrett, *Victorian Multiplot Novel*. For the Marxist tradition, see Cascardi, "Totality and the Novel"; Jameson, introduction to Lukács, *Historical Novel;* and Jay, *Marxism and Totality*. As Choi notes, there is relatively little work explicitly devoted to this widely acknowledged and admired feature of Victorian literary form. She convincingly shows that the novel's approach to material interconnection both informed and was informed by contemporary debates about the spread of contagious disease. This project tries to build on Choi's groundbreaking work by suggesting that a broader category of narratives mapping material interconnection underlay and preceded the more specific stories of epidemic transmission.

13. Boisseron, *Afro-Dog*, 5. But as Boisseron explains, the imperative to counteract such silence does not offer any easy answers about how to deal with issues as thoroughly entwined as race, gender, and species: "One cannot address the entangled oppression of humans, minorities, and animals while doing justice to all groups and individuals equally. . . . [T]he goal of intersectionality should simply be to open our minds to the exponentiality of intersections but with no ambition to address them all" (25–26).

1. APPROACHING ECOLOGY

1. Susan Darwin to Charles Darwin, Shrewsbury, October 15, 1833, in Darwin Correspondence Project, letter no. 219, www.darwinproject.ac.uk/letter/DCP-LETT-219.xml.

2. Darwin, *Journal of Researches*, 136.

3. Susan Darwin to Charles Darwin, Shrewsbury, October 15, 1833.

4. The body of writing on traditional ecological knowledge (TEK) is voluminous—much of it, unfortunately, the work of outsiders providing reports on Native communities. For an influential argument about the value of Indigenous knowledge for modern sustainable practice, see Gadgil, Berkes, and Folke, "Indigenous Knowledge for Biodiversity Conservation." For more recent work that centers Indigenous voices, see the essays in Nelson and Shilling, *Traditional Ecological Knowledge*; and Oakes et al., *Native Voices in Research*. For a lyrical and popular introduction to TEK, see Kimmerer, *Braiding Sweetgrass*.

5. For Malthus's early life and relationship with his father, see Mayhew, *Malthus*, 53–65.

6. Malthus, *Essay*, 3.

7. For the early backlash against Malthus, the Malthusian moment in the 1970s, and recent disavowals, see Mayhew, *Malthus*, chap. 4, chap. 8, and epilogue, respectively. Mayhew's work has helped pave the way for a resurgence of interest in and appreciation for Malthus's work, a welcome reconsideration that includes Bashford and Chaplin, *New Worlds*, and the essays collected in Mayhew, *New Perspectives on Malthus*.

8. Malthus, *Essay*, 33. Subsequent references to the *Essay* will appear parenthetically in the text.

9. On Malthus and the Physiocrats, see Gallagher, *Body Economic*, 43–50.

10. Any money whose purchasing power fluctuates, in other words, would not count as a real fund or store of value.

11. Jonsson, "Island, Nation, Planet," 135.

12. Jonsson, "Island, Nation, Planet," 132.

13. Jonsson, "Island, Nation, Planet," 132.

14. For relatively recent examples, see Foster, *Marx's Ecology*, 87–102; McLane, "Malthus Our Contemporary?," 347–48; and Morton, *Ecological Thought*, 37, 120–21.

15. For Malthus's argument against inequality, see Malthus, *Essay*, 118n1.

16. Gallagher, *Body Economic*, 36, 43.

17. Morton, *Ecological Thought*, 1. Morton sees Malthus as an enemy of ecological thinking, associating him with self-serving austerity measures and a refusal to admit "distributional inequalities"—despite Malthus's direct, repeated attacks on distributional inequality in the first edition of the *Essay*. See Morton, *Ecological Thought*, 120.

18. Alaimo, *Bodily Natures*, 16.

19. Realism is a complex and highly contested term in literature and aesthetics. This working definition is sufficient for the purposes of this literary history, but it does not attempt to encompass the huge range of debates surrounding the topic. For a brief overview of those debates, see R. Williams, "Realism." The version of the term used here loosely follows the widespread understanding of realism best articulated in G. Levine, *Realistic Imagination*, chap. 1.

20. Mayhew argues that Malthus developed a kind of environmental awareness during his travels to Trondheim in 1799 and that these insights found their way into the revised 1803 edition of the *Essay*, offering an early version of environmental

economics. But the essentially ecological foundations of his thinking are present from the very beginning. See Mayhew, *Malthus*, chap. 4.

21. Latour, *Politics of Nature*, 21.

22. Latour, *Politics of Nature*, 54.

23. Leopold, *Sand County Almanac*, 252–53.

24. Leopold, *Sand County Almanac*, 240, 239.

25. On Leopold's life and significance, see Callicott, "Animal Liberation"; Dunlap, *Saving America's Wildlife*, 65–110; and Flader, *Thinking Like a Mountain*.

26. The first edition of the *Essay* includes chapters (omitted from all later editions) explaining how human superiority is not only maintained but facilitated through this apparently painful, nonanthropocentric ordering of the world. See Malthus, *Essay*, 141–58. For the backlash of the Anglican establishment, see Mayhew, *Malthus*, 73–74.

27. Martineau, *Autobiography*, 1:54.

28. For an accessible overview of this particular crisis, see Dick, "On the Financial Crisis, 1825–26."

29. Martineau, *Autobiography*, 1:108.

30. Martineau, *Autobiography*, 1:137.

31. See, for example, Freedgood, "Banishing Panic"; Gaston, "Natural Law"; and David, *Intellectual Women*, 42–43.

32. Quoted in Chapman, *Memorials*, 2:564.

33. Quoted in Chapman, *Memorials*, 2:590. For attacks on Martineau that lampoon her as the grotesque female face of Malthusian ideas, see, for example, "*Illustrations of Political Economy.* Nos. 1–12," 140–43, 151; "Gallery of Literary Characters," 576.

34. In the same year that my argument here first appeared, at least two other scholars simultaneously published pieces on the ecological aspects of Martineau's work—a new turn in readings of Martineau that represents a surprising and gratifying development. For Ayşe Çelikkol, the ecological elements of Martineau (and later, Gaskell) represent a particular fusing of realism with pastoral with politically suspect implications. I am arguing here that these genre categories are not necessary and may be misleading as ways of understanding the ecological components of these works and other forms of realist storytelling. See Çelikkol, "World Ecology." Rebecca Richardson sees Martineau's *Illustrations* as innovative in their tendency to build new ecological insights into a political economy that previously lacked them. My argument is that Martineau was, in fact, expanding upon an existing strain of Malthusian storytelling. See Richardson, "Environmental and Economic Systems."

35. Martineau, "Cinnamon and Pearls," 65.

36. Martineau, "Cinnamon and Pearls," 21.

37. Grove, *Green Imperialism*, 3.

38. Jonsson, *Enlightenment's Frontier*, 4.

39. Jonsson, "Island, Nation, Planet," 130.

40. Martineau, "Weal and Woe," in *Illustrations of Political Economy: Selected Tales*, 96.

41. Martineau, "Weal and Woe," 136.

42. Martineau, "Weal and Woe," 72.

43. While I see this lack of decisive agency as widespread, others have argued that it is confined to imperial and nonwhite subjects and reflects stadial histories of the Scottish Enlightenment. See C. Klaver, "Imperial Economics."

44. Martineau, "Weal and Woe," 94.

45. On the general nineteenth-century interest in rural scenes and changing depictions of them, see O'Gorman, "Rural Scene." For analyses of the function of such scenic descriptions in nineteenth-century literature, see, for example, R. Williams, *Country and the City*, chaps. 12–13; Ebbatson, *Imaginary England*; Helsinger, *Rural Scenes*; and King, *Divine in the Commonplace*.

46. Harriet Martineau, "Sowers Not Reapers," in *Illustrations of Political Economy: Selected Tales*, 297.

47. Martineau, "Sowers Not Reapers," 297.

48. Martineau, "Sowers Not Reapers," 304–5.

49. Martineau, "Sowers Not Reapers," 358.

50. Martineau, "Sowers Not Reapers," 358.

51. Morton, *Ecological Thought*, 28.

52. See Watts, *Reading the Landscape*; Wessels, *Reading the Forested Landscape*; and Macfarlane, *Landmarks*.

53. Caroline Darwin to Charles Darwin, Shrewsbury, October 28, 1833, in Darwin Correspondence Project, letter no. 224, https://www.darwinproject.ac.uk/letter/DCP-LETT-224.xml.

54. As Devin Griffiths has shown in his own examination of the relations between Darwin and Martineau, Darwin scrupulously recorded only his scientific reading habits for future citation and reference, despite strong evidence of a voracious and lifelong appetite for novel reading. See Griffiths, *Age of Analogy*, 230–37.

55. Darwin, *Autobiography*, 113.

56. Darwin, *Autobiography*, 98.

57. Mann and Rogers indicate that the *Illustrations* were responsible for Darwin's reading of Malthus; Ridley suggests it was Darwin's extensive social encounters with Martineau herself. See Mann and Rogers, "Objects and Objectivity," 253n2; and Ridley, "Natural Order of Things."

58. Darwin repeatedly expresses his admiration for Malthus and a despair at the widespread dismissal of Malthus in his letters. See, for example, Charles Darwin to Alfred Russel Wallace, Bromley, July 5, 1866, in Darwin Correspondence Project, letter no. 5145, https://www.darwinproject.ac.uk/letter/?docId=letters/DCP-LETT-5145.xml.

59. Darwin, *Origin*, 61. Subsequent references to the *Origin* will appear parenthetically in the text.

60. This observation would eventually prove pivotal in changing understandings of the environmental history of heathland; see chap. 4 of this book for a more extensive reading.

61. This would be the case, for example, in the Pampas, where Darwin traveled just before his sisters' letters about Martineau arrived.

62. Darwin later learned that other bees also pollinate red clover, a fact that lessens the straightforward clarity of this example of interdependence without undermining Darwin's broader point. The new information led Darwin to pursue a frustrating series of dead-end conjectures that finally left him venting to another naturalist, "I hate myself I hate clover & I hate Bees." See Charles Darwin to John Lubbock, Bournemouth, September 3, 1862, in Darwin Correspondence Project, letter no. 3705, https://www.darwinproject.ac.uk/letter/?docId=letters/DCP-LETT-3705.xml.

63. Quoted in Burdon-Sanderson, "Inaugural Address," 465n1. For the origins, etymology, and orthographic shifts of "ecology," see "ecology" in *Oxford English Dictionary*, 3rd ed., rev. June 2008, www.oed.com/view/Entry/59380; and E. Miller, "Ecology."

64. For the idea of an affordance and the pioneering work applying it to literary forms, see C. Levine, *Forms*, 6–23. Levine's injunction that critics should highlight the sociopolitical consequences of forms as they move between discourses is one of the major inspirations guiding the structure of this book.

65. Although traditionally scholars do not read the *Illustrations* ecocritically, this underlying question—whether the *Illustrations* are damnably compromised by political economic dogma or whether the stories themselves transcend it—is a major fault line in Martineau scholarship. Some critics take her tales at face value as complicit in the oppressive prescriptions of political economy: see Freedgood, "Banishing Panic"; and Gaston, "Natural Law." Others try to redeem the stories by playing up the ways they resist the doctrines they appear to espouse: see Ella Dzelzainis, "Malthus, Women and Fiction," in Mayhew, *New Perspectives on Malthus*, 175; C. Klaver, *A/Moral Economics*, 53–77; Peterson, "From French Revolution"; Sanders, "From 'Political' to 'Human' Economy"; and Webb, *Harriet Martineau*, 118–24.

66. David, *Intellectual Women*, 43.

2. THE PRICE OF INTERCONNECTION

1. For a classic survey of Europe at midcentury, see Robertson, *Revolutions of 1848*. For a recent analysis connecting these movements to newly available forms of writing and print, see Pettitt, *Serial Revolutions 1848*. For an overview of the "hungry forties" and the situation leading up to the Condition of England debate, see Cannadine, *Victorious Century*, chap. 5.

2. Bentham, *Works*, 142.

3. The subgenre is first coherently identified as "the social novel" in Cazamian, *Social Novel in England*. Arnold Kettle coined the term "social-problem novel" in Kettle, "Early Victorian Social-Problem Novel." For a recent and accessible overview, see Simmons Jr., "Industrial and 'Condition of England' Novels." For important analyses of this set of novels as a subgenre of Victorian realism, see Bodenheimer, *Politics of Story*; Gallagher, *Industrial Reformation*; Poovey, *Making a Social Body*, chap. 7; and R. Williams, *Culture and Society*, chap. 4.

4. On the reception and lasting significance of Gaskell's first novel, see Hopkins, *Elizabeth Gaskell*, chap. 4.

5. Hopkins, *Elizabeth Gaskell*, 82.

6. Kingsley, *Alton Locke*, 111; Eliot, *George Eliot Letters*, 2:32. Eliot was a relatively late contributor to this tradition; her novel *Felix Holt, the Radical* did not appear until 1866.

7. Chapman, *Memorials*, 2:165. "Novel of society," as Chapman uses the term, would have referred to high society or the upper classes, not the novel describing interconnected human communities that I am examining here.

8. Courtemanche, "'Naked Truth,'" 391, 403. For an overview of critics who explore the importance of the *Illustrations* as a precursor to the Victorian social novel, see Logan, introduction to Martineau, *Illustrations of Political Economy*, 40–46.

9. Martineau mentions visiting with the Gaskells, for example, in her diary of 1837; see Chapman, *Memorials*, 2:337. For the direct influence of Martineau's tale on Gaskell's novel, see Fryckstedt, "Early Industrial Novel."

10. Nevertheless, as Sukanya Banerjee notes in her excellent analysis of cotton's role in the text, "the novel itself extends an invitation to read denotatively, ecocritically." See Banerjee, "Ecologies of Cotton," 502.

11. Gaskell, *Mary Barton*, 278. Subsequent references to the novel will appear parenthetically in the text.

12. For more on the Victorian funeral industry, see Howarth, "Professionalizing the Funeral Industry."

13. Harriet Martineau, "A Manchester Strike," in *Illustrations of Political Economy*, 216; emphasis in the original.

14. See Galbraith, *History of Economics*, 84–86. This "law" represents a stricter, uncompromising version of the Malthusian point about rising wages and inflation examined as a kind of ecological plotting in chapter 1. Economists continue to debate how strictly Ricardo asserted the law, but its general argument remains a popular justification among market fundamentalists today for socioeconomic inequality and poverty wages.

15. Disraeli, *Sybil*, 65–66.

16. Kingsley, *Alton Locke*, 376. For an in-depth analysis of the way disease narratives figured into Victorian multiplot storytelling, see Choi, *Anonymous Connections*.

17. Dickens, *Bleak House*, 235.

18. See Uglow, *Elizabeth Gaskell*, for Gaskell's early reading and education (chap. 2) and for her father's understanding of, and publications on, political economy (9–11, 52).

19. See, for example, Gradgrind's son named Malthus in Dickens, *Hard Times*, 24; the references to "surplus population" in Dickens, *Christmas Carol*, 14; and Kingsley, *Alton Locke*, 111.

20. Poovey, *Making a Social Body*, 153.

21. On this theory of value and its persistence in microeconomic thought, see Galbraith, *History of Economics*, 65–66.

22. Disraeli, *Sybil*, 115.

23. Kingsley, *Alton Locke*, 258–59.

24. Douglass, *Narrative*, 48–49.

25. See the classic 1933 lectures in Lovejoy, *Great Chain of Being*. For an insightful discussion of Douglass's use of the Great Chain of Being, see Jackson, *Becoming Human*, 48–55. For a broader overview of the place of the Great Chain of Being in discourses of humanity, animality, and Blackness, see Wynter, "Unsettling the Coloniality," 300–301, 313–14.

26. Douglass, *Narrative*, 26.

27. Douglass, "Agriculture and Black Progress," 4:388.

28. Zakiyyah Iman Jackson has recently written of the ways that Douglass was still bound up in this framework of human rights even as he called for more ethical treatment of nonhuman animals. For her, Douglass's refusal to fully challenge this hierarchy—and his subsequent canonization as a key Black thinker of the nineteenth century—is symptomatic of an anti-Black paternalistic system that assigns Black life an ambiguous, fungible status within a racist and anthropocentric moral order. My brief discussion here loosely follows her more complex, in-depth analysis of Douglass's writings on the subject; see Jackson, *Becoming Human*, 45–58. For a different reading of Douglass that understands him as subverting this framework and helping inaugurate a Black tradition that opens up new kinds of moral recognition by embracing animality rather than rejecting it, see Joshua Bennett, *Being Property Once Myself*, 1–4.

29. J. M. I. Klaver, *Apostle of the Flesh*, 504; see 502–7 for the complexity of Kingsley's attitude toward the American Civil War.

30. For a useful online compendium of resources about Douglass's time in Britain and Ireland, see Hannah-Rose Murray, "Frederick Douglass in Britain and Ireland," http://frederickdouglassinbritain.com/.

31. For in-depth discussions of the history of this comparison in Victorian literature and politics, see Gallagher, *Industrial Reformation*, chap. 1; Huzzey, *Freedom Burning*, 84–93.

32. For accessible accounts of the build-up to and achievement of the abolition of the slave trade in British territories, see Thomas, *Slave Trade*, 537–56; Walvin, *England, Slaves, and Freedom*, chap. 5. For the Slavery Abolition Act of 1833 and its gradual emancipation of enslaved peoples through the 1830s, see Walvin, *England, Slaves, and Freedom*, chap. 7.

33. The self-congratulatory vein of British feeling on questions of slavery is mentioned (among other places) in Huzzey, *Freedom Burning*, 80; Walvin, *Making the Black Atlantic*, 164.

34. See, for example, the sketch of such attitudes provided in Walvin, *England, Slaves, and Freedom*, 82–94.

35. For overviews of the relationship between slavery and British prosperity in the seventeenth and eighteenth centuries, see Morgan, *Slavery and the British Empire*, 34–37; Thomas, *Slave Trade*, 243–50, 272–75; Walvin, *Making the Black Atlantic*, chap. 7. For the British textile industry's reliance on cotton grown on slave plantations in the United States through the 1860s and the global supply crisis precipitated by the American Civil War, see Farnie, *English Cotton Industry*, 14–17; Beckert, *Empire of Cotton*, chap. 9.

36. Martineau, "Weal and Woe," 94.

37. In fact, this ecological vision of entanglement and its sociopolitical possibilities closely aligns with the visions that a growing number of race theorists and Black studies scholars have identified as emerging from Black traditions in the Americas and elsewhere—traditions that see the modern emphasis on human rights as a historical trap that tends toward the marginalization and denigration of people of color. See, for example, Joshua Bennett, *Being Property Once Myself*, 4–13; and Jackson, *Becoming Human*, 21–34. Bénédicte Boisseron has argued that the racism of historical and legal equations of Blackness and animality can serve as counterintuitive opportunities for collaboration between Black liberation and animal liberation movements that are often at loggerheads with one another; see Boisseron, *Afro-Dog*. For an important and incisive account of the different versions of humanity that have emerged in the modern era and the ways they have marginalized Black peoples in particular—an account that nevertheless puts its trust in a better and more rigorous mode of naming and studying the human—see the works of Sylvia Wynter, particularly Wynter, "Unsettling the Coloniality"; Wynter and McKittrick, "Unparalleled Catastrophe."

38. Victorian attitudes toward such chimeras—and to mermaids in particular—are analyzed in Ritvo, *Platypus and the Mermaid*, 175–83. For a classic theoretical analysis of chimeras as representative of humans' entanglement with nature and technology, see Haraway, "Cyborg Manifesto." For more recent discussions of chimeras and ecological thought, see Swanson et al., "Bodies Tumbled into Bodies," M1–M12; and Bacchilega and Brown, "Stories We Tell."

39. Kingsley, *Alton Locke*, 340–57.

40. I owe a debt of gratitude to one of my former students, Angela Mauroni, for drawing my attention to the parallels between the mermaid and the working classes. The mermaid's marginalized humanity and her physical location gesture toward colonized and enslaved peoples as well—she is sighted off the coast of New Zealand, after all—but the descriptions in the text emphasize the ways she conforms to the expected appearance and normative standards of an overwhelmingly white British population.

41. For an expansive collection of essays touching on the rise of natural history and its many dimensions, see Jardine, Secord, and Spary, *Cultures of Natural History*.

42. See Secord, "Corresponding Interests"; and Secord, "Science in the Pub."

43. Some critics have gone so far as to argue that Job Legh's outlook and example are closely aligned to Gaskell's realist worldview. See Coriale, "Gaskell's Naturalist"; and Secord, "Elizabeth Gaskell's Social Vision."

44. For the rise of social networks and organizations connecting Victorian naturalists, including those of different classes, see Allen, *Naturalist in Britain*, chap. 8; and Secord, "Corresponding Interests."

45. "Gradgrind, n.," *Oxford English Dictionary Online*, December 2022, https://www.oed.com/view/Entry/80427.

46. Dickens, *Hard Times*, 9.

47. Dickens, *Hard Times*, 9–10.

48. Allen, *Naturalist in Britain*, 70.

49. Wallace, *Annotated Malay Archipelago*, 758.

50. Wallace, *Annotated Malay Archipelago*, 761.

51. For Wallace's early adoption of an environmentalist stance, see Slotten, *Heretic in Darwin's Court*, 352–53; see 365–73 and 436–38 for Wallace's escalating political commitments and eventual adoption of socialism itself. For the lingering environmental shortcomings of Wallace (and Darwin) from a twenty-first-century perspective, see Allen MacDuffie, "Darwin and the Anthropocene," in Griffiths and Kreisel, *After Darwin*, 57–69.

52. Wallace, *Annotated Malay Archipelago*, 101.

53. Wallace, *Annotated Malay Archipelago*, 101. The names Wallace uses for nonwhite peoples have been replaced in brackets to reflect terms now considered both less derogatory and more accurate.

54. Wallace, *Annotated Malay Archipelago*, 117.

55. Wallace, *Annotated Malay Archipelago*, 108.

56. Wallace, *Annotated Malay Archipelago*, 108–13.

57. Wallace, *Annotated Malay Archipelago*, 118.

58. For an unflinching but also evenhanded meditation on the way the Anthropocene's patterns of global exploitation both enabled and compromised the vision of nineteenth-century naturalists—including, centrally, Darwin—see Jesse Oak Taylor, "Darwin after Nature," in Griffiths and Kreisel, *After Darwin*, 19–32.

59. For examples of prominent critics noting these flaws, see Bodenheimer, *Politics of Story*, 5; David, *Fictions of Resolution*, x; Gallagher, *Industrial Reformation*, xii; and R. Williams, *Culture and Society*, 88–91.

60. Kingsley, *Alton Locke*, 365–66. Kingsley himself clearly subscribes to racist hierarchies that refuse full biological humanity to Black people. The resolution to questions of humanity in *Alton Locke* thus manages to be at once racist in its biological categories and universally humanist in its moral ones, backing up the contention of a number of Black studies scholars about the ways that even inclusion in a human rights paradigm has never really guaranteed equal status to Black people. See, for example, Jackson, *Becoming Human*, 20–34; Wynter, "Unsettling the Coloniality," 315–31.

61. For useful overviews of the multiple attacks on faith in the mid-to-late Victorian era, see Lane, *Age of Doubt*; Olson, *Science and Religion*, chaps. 6–8.

3. THE NATURE OF FICTION

1. On the trip to Devon, see Haight, *George Eliot*, 196–210. Many critics have noted the importance of this trip to Eliot's origins as a novelist. See especially the excellent article by Feuerstein, "Realism of Animal Life." The early parts of this chapter broadly align with Feuerstein's perspective, but our accounts part ways regarding the long-term consequences of Eliot's interest in animals.

2. Eliot, *Journals*, 62.

3. On her adoption of the pseudonym, see Haight, *George Eliot*, 219–20. From here forward I will refer to Marian Evans (born Mary Ann Evans) exclusively by her pen name.

4. For just one example of the cherished status of *Middlemarch* today, see Mead, *My Life in Middlemarch*.

5. For Eliot's early career at the *Westminster Review*, see Haight, *George Eliot*, esp. chaps. 4–5.

6. For an extended consideration of Eliot, Chapman, and their radical place in Victorian politics and thought, see Ashton, *142 Strand*, chaps. 3–4.

7. Eliot, "The Natural History of German Life," in *Selected Essays*, 131.

8. Eliot, "Natural History," 111–12.

9. Robbins, *History of Economic Thought*, 173.

10. Malthus, *Principles of Political Economy*, 1:1–2.

11. The term was coined in Schumpeter, *History of Economic Analysis*, 473.

12. Mill, *Elements of Political Economy*, 4.

13. Jevons, *Theory of Political Economy*, 3.

14. For an overview of the value shifts and omissions in political economy as it developed into classical economics in the mid-nineteenth century, see Hadley, Jaffe, and Winter, "Introduction." On the way narrative and metaphor have remained central to economics in the twentieth and twenty-first centuries—and the way economists have taught themselves to overlook the literary forms that structure their quantitative reasoning—see McCloskey, *Rhetoric of Economics*.

15. Eliot, "Natural History," 110.

16. Eliot, "Natural History," 112.

17. See King, *Divine in the Commonplace*.

18. Eliot, *Journals*, 56–57.

19. Eliot, "Natural History," 110.

20. Eliot, "Natural History," 128.

21. Eliot, "Natural History," 130–31.

22. On naturalists depleting the landscape, see Allen, *Naturalist in Britain*, chaps. 6, 7, and 10. On the histories and ideologies behind the aquarium and the Wardian case, see Hamera, *Parlor Ponds;* and Keogh, *Wardian Case*. For studies of the analogies between the worlds of literary realism and the microcosms found in tide pools, aquariums, and Wardian cases, see King, *Divine in the Commonplace*, chap. 3; and Scott, *Chaos and Cosmos*, chap. 5.

23. On the intimacy of natural history with political economy and Wallace's work in Borneo, see chap. 2 of this study.

24. Eliot, *Journals*, 60.

25. Eliot, "Natural History," 110.

26. Jane Welsh Carlyle to Eliot, February 20, 1859, in Eliot, *George Eliot Letters*, 3:18; emphasis in the original.

27. Eliot, *Adam Bede*, 5. Subsequent references to this work will appear parenthetically in the text.

28. Like other foundational terms in literary study, character is notoriously hard to define, but some combination of agency and internal life are the standard criteria. A sensible overview can be found in Forster, *Aspects of the Novel*, chaps. 3–4. For surveys of narratological approaches to character, see Bal, *Narratology*, 104–24;

Chatman, *Story and Discourse*, 107–38. For recent interventions in debates over the nature of character, see Frow, *Character and Person*; and Galvan, "Character."

29. Notably, Woloch does not even consider the possibility of animal characters; he defines characterization only as "the literary representation of imagined human beings." See Woloch, *One vs. the Many*, 14.

30. Keridiana W. Chez argues that this is the role of dogs in Eliot: to act as "prostheses" that help humans learn to better sympathize with one another and that can then disappear. See Chez, *Victorian Dogs, Victorian Men*, chap. 2. What I try to show here is that dogs block or pervert sympathetic energies far more than they enable them.

31. Of Eliot's novels, only *The Mill on the Floss* has received substantive attention from ecocritics thus far. For insightful analyses of the movements of water that underlie the novel's ecological form and its temporality, see McAuley, "George Eliot's Estuarial Form"; and Elizabeth Carolyn Miller, "Fixed Capital and the Flow," in Hensley and Steer, *Ecological Form*, 85–100. While it isn't explicitly ecological, a foundational exploration of the fluid movements that structure *The Mill on the Floss* (and other Victorian novels) is Law, *Social Life of Fluids*, esp. chap. 3. For a general consideration of the power of multiple circulatory systems and metaphors in *Middlemarch* that is never explicitly identified as ecological in structure, see Beer, *Arguing with the Past*, chap. 6.

32. George Levine argues the opposite point—he believes realism strives toward representing exactly this animal otherness. Feuerstein makes a similar case. And James Eli Adams argues that, in *Adam Bede* at least, the silence of animals stands in for the ideal, wordless realm where realist moral sympathy begins. But the argument of this chapter is that even though Eliot sees and admires sympathy with the nonhuman as an imaginative endpoint, she decides early in her career that it is unattainable for the time being and creates more problems than it solves. See G. Levine, *Realism, Ethics, and Secularism*, chap. 9; Feuerstein, "Realism of Animal Life"; and Adams, "Gyp's Tale."

33. Eliot, "Natural History," 111.

34. On Eliot's evolving religious beliefs and their impact on her fiction, see Qualls, "George Eliot and Religion."

35. For influential feminist readings of Maggie, see, for example, Auerbach, "Power of Hunger"; David, *Intellectual Women*, chap. 12; Fraiman, *Unbecoming Women*, chap. 5; and Jacobus, "Question of Language."

36. This is hardly an original point. But critics typically treat it either as further reinforcement of Maggie's refusal to conform or as a sign of some early Darwinian influence on the novel. For examples of each respective reading, see Auerbach, "Power of Hunger"; and Corbett, "'Crossing o' Breeds.'"

37. Eliot, *Mill on the Floss*, 12, 13. Subsequent references to this work will appear parenthetically in the text.

38. On figures of "the gypsy" in British literature, see Nord, *Gypsies*.

39. This case for the cultural function of the bildungsroman—and the particular Englishness of the compromising happy ending—was made most famously in Moretti, *Ways of the World*. Important treatments of *The Mill on the Floss* as a

bildungsroman include Esty, *Unseasonable Youth*, chap. 2; Fraiman, *Unbecoming Women*, chap. 5; Hensley, *Forms of Empire*, chap. 1; and Hildebrand, "Environmental Desire." None of them deals at length with the question of Maggie's animality and its significance for her maturity.

40. Feder, *Ecocriticism*, 19. Feder does not include Eliot's novel in her analysis.

41. Hensley, "Database," 119, and see 128–30 for a brief overview of the long-standing controversy over the ending.

42. Coleridge, *Biographia Literaria*, 2:6.

43. Forster, *Aspects of the Novel*, 86.

44. Lukács, "Narrate or Describe?," 127, 131, 126.

45. Lukács, "Narrate or Describe?," 127.

46. Lukács, "Narrate or Describe?," 131.

47. For description as a way of building such flat, autonomous settings, see Lukács, "Narrate or Describe?," 113–16. A useful history of the marginalization of description is found in Hamon, "Rhetorical Status." Description and the scenery it creates remain an ongoing problem for the study of narrative known as narratology; for representative treatments of its odd place in the field, see Bal, *Narratology*, 26–27; Chatman, *Story and Discourse*, 264.

48. Morton, *Ecology Without Nature*, 32.

49. Lukács, "Narrate or Describe?," 125, 137, 139.

50. Despite noting Eliot's departure from Darwin, Hildebrand sees this feature of Eliot's writing as a promising ecological development. By contrast, I try to show here that it actually marks the end of Eliot's serious engagement with the intellectual tradition that gave rise to ecology. See Hildebrand, "Environmental Desire"; Hildebrand, "*Middlemarch*'s Medium." For the origins of this shift and its effects on the history of science, see Pearce, "From 'Circumstances' to 'Environment.'"

51. Eliot, "Modern Painters, Vol. III," in *Selected Essays*, 370.

52. Eliot, "Modern Painters, Vol. III," 371; emphasis in the original.

53. Maggie first feels this when her father stops prioritizing her and focuses instead on Tom's prospects as a young businessman.

54. See Hildebrand, "Environmental Desire." Hildebrand sees this as a positive, preferable alternative to the more commercial and possessive kinds of desire prevalent in the text and as a forerunner of modern environmental thought. But I try to show here how it represents Eliot's ambivalent but decisive break from an ecological mode committed to detailing how interspecies relationships work.

55. Nina Auerbach has already observed and lamented Eliot's lack of investment in nonhuman individuality. See Auerbach, "Dorothea's Lost Dog."

56. Fetch goes from female to male between chapters 12 and 13. See Eliot, *Daniel Deronda*, 109, 117. Osborne notes the shift in his edition of Eliot, *Daniel Deronda*, 751n3.

57. Eliot, *Middlemarch*, 141.

58. Eliot, *Middlemarch*, 194.

59. Despite the paucity of actual paws in Eliot's later work, some critics have tried to reclaim Eliot's animal metaphors as indicative of her attunement with Darwin or her interest in a model of interspecies sympathy. See, for example, Kingstone,

"Human-Animal Elision"; Kreilkamp, *Minor Creatures*, chap. 4; and Pielak, "Hunting Gwendolen."

60. James, "Preface to *Roderick Hudson*," 1041.

61. Trollope, *Autobiography*, 317; he describes his daily writing quota on 236–38.

62. Eliot, *Daniel Deronda*, 369.

63. Eliot, *Daniel Deronda*, 379.

64. Eliot, *Daniel Deronda*, 380.

65. In that sense, the complex negotiation of character and setting in Eliot complicates the history of the novel's engagement with the environment laid out recently by Amitav Ghosh. Like Eliot and James, Ghosh is a novelist interested in the way the novel draws boundaries and makes exclusions, but he sees the realist novel in a more sinister light as a monolithic tool for suppressing both environmental and geopolitical interconnectedness—and thus as a technology that was instrumental to shaping our disastrous ecological present. The account I have been laying out here suggests that realist storytelling was also instrumental in the rise of the very same ecologically informed perspectives now used to critique it. See Ghosh, *Great Derangement*, 1–84.

66. In her journal, Eliot first describes Darwin's book as "not impressive, from want of luminous and orderly presentation." See Eliot, *Journals*, 82. Her letters are more positive, but they treat Darwin's work as an admirable synthesis and summary of evolutionary ideas rather than revolutionary in its own right. See Eliot, *George Eliot Letters*, 3:214.

4. THE STORY OF ECOLOGY

1. For a "digest biography" of Gissing's life, see Tyndall, *Born Exile*, 9–18.

2. A good sense of the relations between the two men can be gathered from Gissing, *London*.

3. Gissing to Algernon Gissing, Epsom, September 22, 1895, in Gissing, *Collected Letters*, 6:28.

4. The phrase "landscape novelist" comes from Irwin, *Reading Hardy's Landscapes*, 4. One of the earliest examples of Hardy's canonization as a novelist deeply attuned to nature is found in Lea, *Thomas Hardy's Wessex*, 67. For a sampling of recent criticism that casts Hardy's novels as sources of ecological insight or models of ecological writing, see W. Cohen, "Arborealities"; Feuerstein, "Seeing Animals"; Kerridge, "Ecological Hardy"; and E. Miller, "Dendrography and Ecological Realism." While George Levine never explicitly uses the word "ecology" in *Reading Thomas Hardy*, he similarly advocates for Hardy as a writer who can teach readers humility in relation to nature alongside an understanding of nature's interconnectedness.

5. Millgate, *Thomas Hardy*, 244.

6. Hardy attended Darwin's funeral in 1882 and described himself as "among the earliest acclaimers of *The Origin of the Species*" in Hardy, *Life and Work*, 158. For a classic study that insists on the important relationship between Darwin and Hardy, see Beer, *Darwin's Plots*, chap. 8. For a recent and comprehensive survey, see G. Levine, *Reading Thomas Hardy*.

7. For the novel as a significant milestone in Hardy's career, in particular with regard to his representations of nature and his existential pessimism, see Langbaum, *Thomas Hardy*, 100; G. Levine, *Reading Thomas Hardy*, 116; Gregor, *Great Web*, 110; Howe, "Return of the Native," 13–14.

8. Hardy's proximity to Eliot is apparent in his autobiography, where he notes that his early anonymous stories sometimes resulted in critics guessing that Eliot was their author—much as Gaskell's first novel was attributed to Martineau. See Hardy, *Life and Work*, 100.

9. On the formulation of the Wessex edition that included this grouping, see Millgate, *Thomas Hardy*, 436–38.

10. For the shifting meanings of environment and environmentalism, see Pearce, "From 'Circumstances' to 'Environment'"; and Winter, *Secure from Rash Assault*, 19–20.

11. The rise of environmentalism as a coherent postwar movement—first in the United States, then across the globe—is thoroughly documented in Rome, *Bulldozer in the Countryside*; Rome, *Genius of Earth Day*; and Sellers, *Crabgrass Crucible*.

12. For examples of critics who unapologetically call Egdon Heath a character, see Fleishman, *Fiction*, 110–22; Langbaum, *Thomas Hardy*, 64. For critics who acknowledge this tradition of treating Egdon Heath as a character but go on to contest it, see Gregor, *Great Web*, 81–82; J. Hillis Miller, *Thomas Hardy*, 91; and M. Williams, *Thomas Hardy*, 136–45.

13. Hardy, *Return of the Native*, 9. Subsequent references to this work will appear parenthetically in the text.

14. These kinds of descriptions led early admirers to embrace the novel as an example of humanity's harmony with the landscape, as in Lea, *Thomas Hardy's Wessex*, 67. Early environmentally attuned readings of Hardy took a similarly laudatory approach; see Enstice, *Thomas Hardy*, 81–84. Despite their more sophisticated use of theory, recent readings continue to treat such blurrings of human and natural traits as profound marks of Hardy's awareness of continuities between the human and the nonhuman. See, for example, W. Cohen, "Faciality and Sensation." For an incisive critique of this infatuation with blurriness, see Cohn, "'No Insignificant Creature.'"

15. Emerson, "Poet," 22.

16. Carroll, *Alice's Adventures in Wonderland*, 125, 136. The question of whether metaphors ever really "die"—and how they continue to shape language—is interestingly discussed in Lakoff and Johnson, *Metaphors We Live By*; McCloskey, *Rhetoric of Economics*; and Pawelec, "Death of Metaphor."

17. Jakobson and Halle, *Fundamentals of Language*, 90–96.

18. For alternative readings that see Egdon Heath resisting privatization and the violation of traditional common rights rather than humanity as a whole, see Lamb, "Storied Matter"; and Lesjak, *Afterlife of Enclosure*, 148–54.

19. Ngai, *Our Aesthetic Categories*, 92. For a recent engagement with Mori's concept that includes an accessible English translation of the Japanese original, see MacDorman, "Androids." Whereas I see the uncanny valley as the product of a unique tension that plays out over time between personification and objectification, others

have argued that it is a dynamic underlying all personification; see Ngai, *Our Aesthetic Categories*, 91–94.

20. Beer, *Darwin's Plots*, 229.

21. G. Levine, *Reading Thomas Hardy*, 119–20.

22. Hardy, *Tess of the d'Urbervilles*, 191.

23. Hardy, *Tess of the d'Urbervilles*, 115.

24. Hardy, *Jude the Obscure*, 38–39.

25. Timothy Morton convincingly argues that the uncanny valley is the feeling most directly opposed to ecological understanding. The experience is a symptom of the "hard separation of things into subjects and objects" that makes recognizing interconnection impossible. See Morton, *Dark Ecology*, 137.

26. Rackham, *History of the Countryside*, 303.

27. Parry, *Living Landscapes*, 32.

28. Tilley, *Landscape*, 370.

29. Darwin, *On the Origin of Species*, 65. Subsequent references to this work will appear parenthetically in the text. On Darwin, Haeckel, and the origins of the word "ecology," see E. Miller, "Ecology"; see also chap. 1 of this study.

30. Paradis, "Darwin and Landscape," 86.

31. For a historical and philosophical overview of the widespread use of anthropomorphism in the arts and sciences, see Daston and Mitman, "Introduction." For a comprehensive explanation and defense of anthropomorphism from a scientific perspective, see de Waal, "Anthropomorphism and Anthropodenial." For recent arguments in favor of anthropomorphism in political and cultural studies, see Jane Bennett, *Vibrant Matter*, 98–100; Griffiths, *Age of Analogy*, 245–55; and Latour, *Facing Gaia*, 57–58, 109–10.

32. For the gradual scientific confirmation of Darwin's theory, see Chadwick, *In Search of Heathland*, 38–42; and Gimingham, *Introduction to Heathland Ecology*, 13–16. For the role of grazing animals in the maintenance of heathland, see Gimingham, *Introduction to Heathland Ecology*, chap. 6; Howkins, *Heathers and Heathlands*, 28–38; and Parry, *Living Landscapes*, 60–61. On the colonization of heaths by Scots pine and European white birch, see Gimingham, *Introduction to Heathland Ecology*, 45–46; and Parry, *Living Landscapes*, 118–19.

33. Beer, *Darwin's Plots*, xxv.

34. For the groundbreaking analysis of narrative and metaphors in Darwin's writing, see Beer, *Darwin's Plots*, chaps. 3–4. For an important recent account that still treats Darwin's narratives almost exclusively in relation to natural selection, see G. Levine, *Darwin the Writer*, chaps. 1–3. Despite an initial focus on landscape, James Krasner's visual approach also prioritizes Darwin's techniques of imaging evolution; see Krasner, *Entangled Eye*, chap. 1.

35. Griffiths, *Age of Analogy*, 238.

36. Darwin, *Formation of Vegetable Mould*, 2, 6. Darwin mentions the 1842 foundations of his work on worms in *Autobiography*, 82.

37. For traditional uses of turves, heather, and furze, see Howkins, *Heathers and Heathlands*, 11–17; and Parry, *Living Landscapes*, 62–70. For heathland wildlife and food

webs, see Gimingham, *Introduction to Heathland Ecology*, chap. 5; Parry, *Living Landscapes*, chap. 4; and—for a special focus on the Dorset heathlands Hardy fictionalized—Chadwick, *In Search of Heathland*, 166–83.

38. Benjamin Morgan, "Scale as Form," in Menely and Taylor, *Anthropocene Reading*, 132. On Hardy and deep time, see also Beer, *Darwin's Plots*, chap. 8; Buckland, "Physics, Geology, Astronomy"; and Padian, "Evolution and Deep Time." For a recent and productive reconsideration of these scalar leaps as matters of genre, see Aaron Rosenberg, "'Infinitesimal Lives,'" in Hensley and Steer, *Ecological Form*, 182–99.

39. J. Cohen, *Stone*, 83.

40. On the loss of heathland because of shifts away from traditional folkways and common rights, see Chadwick, *In Search of Heathland*, 48–53; Howkins, *Heathers and Heathlands*, 80–82; Parry, *Living Landscapes*, 49–53; and Rackham, *History of the Countryside*, 297. On government expropriation of heathland for military purposes, see Parry, *Living Landscapes*, 54–55. For the Forestry Commission's encouragement of heathland afforestation, see Howkins, *Heathers and Heathlands*, 83–84; Parry, *Living Landscapes*, 57; and Rackham, *History of the Countryside*, 297. For figures on the rate and percentage of decline of Dorset heathland, see Rackham, *History of the Countryside*, 302. For information about unique heathland fauna and their decline in Britain, see Chadwick, *In Search of Heathland*, 166–79; and Parry, *Living Landscapes*, chap. 4. For an early European experiment that influenced English attempts at "reclamation" of heath "wastelands" through cultivation and afforestation, see Winter, *Secure from Rash Assault*, 155–58.

41. For an invaluable exploration of the genres used to represent environmental loss, see Heise, *Imagining Extinction*, chaps. 1–2.

42. Tsing, "Arts of Inclusion," 198, 199.

43. On the rise of objectivity as a goal in scientific discourse, see, for example, Daston and Galison, *Objectivity*. On the rise of similarly quantitative and objective ideals in the social sciences, see Porter, *Trust in Numbers*.

44. On ecology and wicked problems, see Defries and Nagendra, "Ecosystem Management."

CONCLUSION

1. For Elton's biography and background, see Southwood and Clarke, "Charles Sutherland Elton."

2. McIntosh, *Background of Ecology*, 89, 91; Crowcroft, *Elton's Ecologists*, xvi. On Warming and his influence, see Worster, *Nature's Economy*, 198–204. For early American theories of succession, see Hagen, *Entangled Bank*, chap. 2. For the British response devoted to less wild and pristine plant ecosystems, see Worster, *Nature's Economy*, 239–42.

3. Elton, *Animal Ecology*, 1. Subsequent references to this work will appear parenthetically in the text.

4. On Elton's reading of Carr-Saunders, see Crowcroft, *Elton's Ecologists*, 4.

5. Huxley, "Editor's Introduction," in Elton, *Animal Ecology*, xiv.

6. A concise overview of Elton's contributions to the crystallizing field is found in Worster, *Nature's Economy*, 294–301. His specific redefinition of the idea of an ecological niche is explained in Crowcroft, *Elton's Ecologists*, xiii, xvii. Elton's insights drew on the newly codified idea of energy emerging in the field of thermodynamics—a science that was still young and controversial in the 1800s. For the tense relations between thermodynamics and ecology in the nineteenth century, see Gold, *Energy, Ecocriticism*, chap. 3.

7. See, for example, Elton, *Animal Ecology*, 107.

8. Worster, *Nature's Economy*, 313.

9. For Elton's reliance on industry and economic applications for funding, see Crowcroft, *Elton's Ecologists*, 5, 28–46.

10. See also Elton, *Animal Ecology*, vii–viii, 177, 189.

11. Huxley, "Editor's Introduction," xiv.

12. Huxley, "Editor's Introduction," xiv, xv.

13. The battle over Thirlmere in the Lake District is thoroughly documented in Ritvo, *Dawn of Green*. The classic account of such disputes in the American context is found in Nash, *Wilderness*, 129–30, 161–81.

14. On their friendship and its impacts on Leopold's ecological thought, see Flader and Callicott, *River*, 6, 14, 224–25; and Warren, *Aldo Leopold's Odyssey*, 136–41, 188–95.

15. Leopold, *Sand County Almanac*, 246, 247. This same form of reasoning continues to be employed in environmental education today.

16. Leopold, *Sand County Almanac*, 247.

17. See, most pertinently, his reflections on the ecological and moral consequences of his job killing wolves in Leopold, *Sand County Almanac*, 137–41.

18. Leopold, *Sand County Almanac*, 262.

19. Traditionally, historians have followed Carson herself in tracing her interest in DDT to her correspondence with women whose yards had been sprayed in 1958. An in-depth account can be found in Lear, *Rachel Carson*, 313–22. But as Lear herself discovered, Carson began pitching the possibility of a piece on DDT as early as 1945, following on earlier concerns about arsenic contamination from the 1930s. On these prior concerns, see Lear, *Rachel Carson*, 312. For their link to nuclear testing, see Lytle, *Gentle Subversive*, 120–21. For Carson's early encounter with Elton, see Lytle, *Gentle Subversive*, 201–3.

20. See Lytle, *Gentle Subversive*, 205–17, for the events between the book's publication and the DDT ban, and see 219–28 for the memorialization of Carson as a radical—both by her more extreme proponents and by her detractors.

21. Carson, *Silent Spring*, 74, 75.

22. Leopold, *Sand County Almanac*, 262.

23. Leopold, *Sand County Almanac*, 251.

24. Leopold, *Sand County Almanac*, xviii. Leopold's family has engaged in some reorganization and retitling in the years since its first publication, including adding several essays and expanding the book into four sections. For details on those changes, see Leopold, *Sand County Almanac*, xiii.

25. Leopold, *Sand County Almanac*, 111.

26. On it-narratives, see Bellamy, "It-Narrators and Circulation." While they might seem like natural illustrations of interdependence and thus in some sense precursors to ecological storytelling, it-narratives were typically employed for comic purposes only—they offered a chance to run through a gallery of amusing stereotypes of figures drawn from all levels of society.

27. Carson, *Silent Spring*, 2, 3.

28. Carson, *Silent Spring*, 3.

29. Carson, *Silent Spring*, 105–9.

30. Carson, *Silent Spring*, 293.

31. Carson, *Silent Spring*, 109.

32. The exact cause of Carson's cancer is, of course, unknowable. For her diagnosis and eventual death, see Lear, *Rachel Carson*, 364–69, 476–80.

33. A classic survey of the modernist novel's techniques of representing consciousness is Humphrey, *Stream of Consciousness*. For an overview of modernism's combined turns inward to the mind and to the novel itself, see Fletcher and Bradbury, "Introverted Novel." The once-dominant story of modernism as an "inward turn" is increasingly disputed by scholars of modernism. See, for example, Gang, "Mindless Modernism"; and Jay, "Modernism." But what scholars dispute now is what kind of psychological model modernists followed or how this inward turn affected the politics of the movement. Modernists' waning interest in realist social panoramas and their narrowing of focus to more bounded subjects—whether the mind or the artistic form of the novel itself—are not really up for debate.

34. Hovanec, *Animal Subjects*, 196.

35. For examples of green readings of modernism, see McCarthy, *Green Modernism;* and Sultzbach, *Ecocriticism*. These analyses effectively overturn older paradigms of ignoring the nonhuman environments in modernism, showing that some entities representing nature do exist in these texts and do play a role in the works' political and artistic outlooks. It would be a stretch, however, to call such entities proof of anything like concerted ecological thought, much less ecological ethics. For an analysis of the rejection of ecological ideas in modernist poetry, see Schuster, *Ecology of Modernism*. On the crucial distinction between reading something ecologically and reading a text that is itself ecological, see Gold, *Energy, Ecocriticism*, chaps. 1–2. The big difference between such exercises in ecological reading and this study of ecological genealogy is that I begin with the observation that the literary tradition under examination is fundamentally *not* ecological, which explains the historical lack of ecological responses and consequences of the work. What I have tried to show here is why even these Victorian works explicitly devoted to thinking through interdependence never achieved anything like a recognizable ecological ethic.

36. I have written about this aspect of weird fiction. See John Miller, "Weird Beyond Description." For a recent accessible argument that weird fiction offers a promising ecological mode, see Wilk, *Death by Landscape*.

37. On animal agents in detective fiction, see Kreilkamp, *Minor Creatures*, chap. 6.

38. The import and fate of such animal stories is usefully summarized in Lutts, "Animals and Ideas." Until recently, the same lack of seriousness associated with the

nonhuman was a problem for scholarship as well. Una Chaudhuri has described what she calls "the laugh test," where those seriously engaged with nonhuman life test out the openness of others by seeing whether they can even take the subject seriously. See Chaudhuri, "(De)Facing the Animals."

39. Baldwin, "Somewhere a Dog Barked."

40. On the tendency of environmental storytelling to turn to mourning, loss, and apocalypticism, see Heise, *Imagining Extinction*, chaps. 1–2. For accessible examples of species recoveries that complicate stories of loss and reframe the roles humans must play in twenty-first-century ecosystems, see Preston, *Tenacious Beasts*. For examples of recent critiques of the stories environmentalists tell about nonnative and introduced species, see Kim, *Dangerous Crossings*, 150–58; and Fugate and Miller, "Shakespeare's Starlings." For the failure of Western environmentalists and scientists to acknowledge positive, reciprocal relationships between humanity and the natural world, see Kimmerer, *Braiding Sweetgrass*, 5–10.

41. For a brief and excellent introduction to these problems with economic modeling and their failure to capture the chains of human and nonhuman agencies that make up the world, see Kapp, "Environmental Disruption."

42. For examples of how ecological economics critiques neoclassical truisms, see Kapp, "Environmental Disruption"; Daly, *Ecological Economics*. For a classic study that critiqued the expected utility theory of economic decision-making and helped catalyze the rise of behavioral economics, see Kahneman and Tversky, "Prospect Theory."

43. Wynter and McKittrick, "Unparalleled Catastrophe," 66, 58; emphasis in the original.

44. Kim, *Dangerous Crossings*, 19; emphasis in the original.

45. There are at least some signs that this process of reconsidering the exclusion of the nonhuman may be underway, and ecological plots that reincorporate plants and animals may be entering mainstream fiction once again. Wai Chee Dimock has recently argued that this expansive vision beyond the human, which she sees as being aligned with the novel's dormant epic impulse, has been reemerging and may prove crucial to the novel's survival as a useful social form. See Dimock, "The Survival of the Unfit," in Griffiths and Kreisel, *After Darwin*, 121–33. There are also important movements among ecocritics to become more seriously interdisciplinary in the work that they do. Empirical ecocritics, for example, insist that empirical studies are vital to grounding claims about a story's environmental potential and significance in a more rigorous framework; see Schneider-Mayerson et al., *Empirical Ecocriticism*. My critique of the ecocritical valorization of Thomas Hardy and others is intended as part of this call for greater interdisciplinarity: I am arguing that it is crucial to be conversant in ecology and environmental history to make a compelling claim about the significance of a writer's ideas to ecological understanding.

Bibliography

Adams, James Eli. "Gyp's Tale: On Sympathy, Silence, and Realism in *Adam Bede*." *Dickens Studies Annual* 20 (1991): 227–42.

Alaimo, Stacy. *Bodily Natures: Science, Environment, and the Material Self.* Bloomington: Indiana University Press, 2010.

Allen, David Elliston. *The Naturalist in Britain: A Social History.* 2nd ed. Princeton, NJ: Princeton University Press, 1994.

Almquist, Julka, and Julia Lupton. "Affording Meaning: Design-Oriented Research from the Humanities and Social Sciences." *Design Issues* 26, no. 1 (2010): 3–14.

Ashton, Rosemary. *142 Strand: A Radical Address in Victorian London.* London: Chatto and Windus, 2006.

Auerbach, Nina. "Dorothea's Lost Dog." In *Middlemarch in the Twenty-First Century*, edited by Karen Chase, 87–105. New York: Oxford University Press, 2006.

———. "The Power of Hunger: Demonism and Maggie Tulliver." *Nineteenth-Century Fiction* 30, no. 2 (1975): 150–71.

Bacchilega, Christina, and Marie Alohalani Brown. "The Stories We Tell about Mermaids and Other Water Sprites." In *The Penguin Book of Mermaids,* edited by Christina Bacchilega and Marie Alohalani Brown, ix–xxix. New York: Penguin, 2019.

Bal, Mieke. *Narratology: Introduction to the Theory of Narrative.* 4th ed. Toronto: University of Toronto Press, 2017.

Baldwin, Rosecrans. "Somewhere a Dog Barked." *Slate,* June 17, 2010. https://slate.com/culture/2010/06/pick-up-just-about-any-novel-and-you-ll-find-the-phrase-somewhere-a-dog-barked.html.

Banerjee, Sukanya. "Ecologies of Cotton." *Nineteenth-Century Contexts* 42, no. 5 (2020): 493–507.

Barton, Ruth. "'Men of Science': Language, Identity, and Professionalization in the Mid-Victorian Scientific Community." *History of Science* 41, no. 1 (2003): 73–119.

Bashford, Alison, and Joyce E. Chaplin. *The New Worlds of Thomas Robert Malthus: Rereading the "Principle of Population."* Princeton, NJ: Princeton University Press, 2016.

Bate, Jonathan. *Romantic Ecology: Wordsworth and the Environmental Tradition.* London: Routledge, 1991.

———. *The Song of the Earth.* Cambridge, MA: Harvard University Press, 2002.

Beckert, Sven. *Empire of Cotton: A Global History.* New York: Alfred A. Knopf, 2015.
Beer, Gillian. *Arguing with the Past: Essays in Narrative Form from Woolf to Sidney.* London: Routledge, 1989.
———. *Darwin's Plots: Evolutionary Narrative in Darwin, George Eliot and Nineteenth-Century Fiction.* 2nd ed. New York: Cambridge University Press, 2000.
Bellamy, Liz. "It-Narrators and Circulation: Defining a Subgenre." In *The Secret Life of Things: Animals, Objects, and It-Narratives in Eighteenth-Century England,* edited by Mark Blackwell, 117–46. Lewisburg, PA: Bucknell University Press, 2007.
Bennett, Jane. *Vibrant Matter: A Political Ecology of Things.* Durham, NC: Duke University Press, 2010.
Bennett, Joshua. *Being Property Once Myself: Blackness and the End of Man.* Cambridge, MA: Harvard University Press, 2020.
Bentham, Jeremy. *The Works of Jeremy Bentham.* Vol. 10. Edited by John Bowring. Edinburgh: William Tait, 1842.
Bodenheimer, Rosemary. *The Politics of Story in Victorian Social Fiction.* Ithaca, NY: Cornell University Press, 1988.
Boisseron, Bénédicte. *Afro-Dog: Blackness and the Animal Question.* New York: Columbia University Press, 2018.
Brontë, Charlotte. *Jane Eyre.* Edited by Margaret Smith. New York: Oxford University Press, 2019.
Brown, Daniel. *The Poetry of Victorian Scientists: Style, Sense and Nonsense.* New York: Cambridge University Press, 2013.
Buckland, Adelene. "Physics, Geology, Astronomy." In *Thomas Hardy in Context,* edited by Phillip Mallett, 242–52. Cambridge: Cambridge University Press, 2013.
Buell, Lawrence. *The Environmental Imagination: Thoreau, Nature Writing, and the Formation of American Culture.* Cambridge, MA: Harvard University Press, 1995.
———. *The Future of Environmental Criticism: Environmental Crisis and the Literary Imagination.* Malden, MA: Blackwell, 2005.
Burdon-Sanderson, J. S. "Inaugural Address." *Nature* 48, no. 1246 (1893): 464–72.
Callicott, J. Baird. "Animal Liberation: A Triangular Affair?" In *The Animal Rights / Environmental Ethics Debate,* edited by Eugene C. Hargrove, 37–69. Albany: State University of New York Press, 1992.
Cannadine, David. *Victorious Century: The United Kingdom, 1800–1906.* New York: Viking, 2017.
Carroll, Lewis. *Alice's Adventures in Wonderland and Through the Looking-Glass and What Alice Found There.* 150th anniversary ed. New York: Penguin, 2015.
Carson, Rachel. *Silent Spring.* Boston: Mariner Books, 2002.
Cascardi, A. J. "Totality and the Novel." *New Literary History* 23, no. 3 (1992): 607–27.
Cazamian, Louis. *The Social Novel in England, 1830–1850: Dickens, Disraeli, Mrs. Gaskell, Kingsley.* Translated by Martin Fido. London: Routledge, 1973.
Çelikkol, Ayşe. "World Ecology in Martineau's and Gaskell's Colonial Pastorals." *Journal of Victorian Culture* 25, no. 1 (2020): 110–25.
Chadwick, Lee. *In Search of Heathland.* London: Dennis Dobson, 1982.

Chapman, Maria Weston. *Memorials of Harriet Martineau*. In Harriet Martineau, *Autobiography and Memorials*, edited by Maria Weston Chapman, 2:133–596. Boston: James R. Osgood, 1877.

Chatman, Seymour. *Story and Discourse: Narrative Structure in Fiction and Film*. Ithaca, NY: Cornell University Press, 1980.

Chaudhuri, Una. "(De)Facing the Animals: Zooësis and Performance." *TDR: The Drama Review* 51, no. 1 (2007): 8–20.

Chez, Keridiana W. *Victorian Dogs, Victorian Men: Affect and Animals in Nineteenth-Century Literature and Culture*. Columbus: Ohio State University Press, 2017.

Choi, Tina Young. *Anonymous Connections: The Body and Narratives of the Social in Victorian Britain*. Ann Arbor: University of Michigan Press, 2015.

Clark, Timothy. *The Cambridge Introduction to Literature and the Environment*. New York: Cambridge University Press, 2011.

Cohen, Jeffrey Jerome. *Stone: An Ecology of the Inhuman*. Minneapolis: University of Minnesota Press, 2015.

Cohen, William A. "Arborealities: The Tactile Ecology of Hardy's *Woodlanders*." *19: Interdisciplinary Studies in the Long Nineteenth Century* 19 (2014). doi.org/10.16995/ntn.690.

———. "Faciality and Sensation in Hardy's *Return of the Native*." *PMLA: Proceedings of the Modern Language Association* 121, no. 2 (2006): 437–52.

Cohn, Elisha. "'No Insignificant Creature': Thomas Hardy's Ethical Turn." *Nineteenth-Century Literature* 64, no. 4 (2010): 494–520.

Coleridge, Samuel Taylor. *Biographia Literaria*. 2 vols. Edited by James Engell and W. Jackson Bate. Princeton, NJ: Princeton University Press, 1983.

Commoner, Barry. *The Closing Circle: Nature, Man, and Technology*. New York: Alfred A. Knopf, 1971.

Corbett, Mary Jean. "'The Crossing o' Breeds' in *The Mill on the Floss*." In *Victorian Animal Dreams: Representations of Animals in Victorian Literature and Culture*, edited by Deborah Denenholz Morse and Martin A. Danahay, 121–44. Burlington, VT: Ashgate, 2007.

Coriale, Danielle. "Gaskell's Naturalist." *Nineteenth-Century Literature* 63, no. 3 (2008): 346–65. https://doi.org/10.1525/ncl.2008.63.3.346.

Courtemanche, Eleanor. "'Naked Truth Is the Best Eloquence': Martineau, Dickens, and the Moral Science of Realism." *ELH* 73, no. 2 (2006): 383–407. https://doi.org/10.1353/elh.2006.0014.

Cowles, Henry. *The Scientific Method: An Evolution of Thinking from Darwin to Dewey*. Cambridge, MA: Harvard University Press, 2020.

Crowcroft, Peter. *Elton's Ecologists: A History of the Bureau of Animal Population*. Chicago: University of Chicago Press, 1991.

Csiszar, Alex. *The Scientific Journal: Authorship and the Politics of Knowledge in the Nineteenth Century*. Chicago: University of Chicago Press, 2018.

Daly, Herman E. *Ecological Economics and Sustainable Development: Selected Essays of Herman Daly*. Cheltenham, UK: Edward Elgar, 2007.

Darwin, Charles. *The Autobiography of Charles Darwin, 1809–1882*. Edited by Nora Barlow. New York: W. W. Norton, 2005.

———. *The Formation of Vegetable Mould, through the Action of Worms, with Observations on Their Habits*. New York: D. Appleton, 1883.

———. *Journal of Researches into the Natural History and Geology of the Countries Visited During the Voyage of H.M.S. Beagle*. 2nd ed. Cambridge: Cambridge University Press, 2011.

———. *On the Origin of Species*. Edited by Gillian Beer. Oxford: Oxford University Press, 1996.

Darwin Correspondence Project. Directed by James A. Secord and Alison Pearn. www.darwinproject.ac.uk.

Daston, Lorraine. "The Academies and the Unity of Knowledge: The Disciplining of the Disciplines." *Differences: A Journal of Feminist Cultural Studies* 10, no. 2 (1998): 67–86.

Daston, Lorraine, and Peter Galison. *Objectivity*. New York: Zone Books, 2010.

Daston, Lorraine, and Gregg Mitman. "Introduction: The How and Why of Thinking with Animals." In *Thinking with Animals: New Perspectives on Anthropomorphism*, edited by Lorraine Daston and Gregg Mitman, 1–14. New York: Columbia University Press, 2005.

David, Deirdre. *Fictions of Resolution in Three Victorian Novels: "North and South," "Our Mutual Friend," "Daniel Deronda."* New York: Columbia University Press, 1981.

———. *Intellectual Women and Victorian Patriarchy: Harriet Martineau, Elizabeth Barrett Browning, George Eliot*. Ithaca, NY: Cornell University Press, 1987.

Defries, Ruth, and Harini Nagendra. "Ecosystem Management as a Wicked Problem." *Science* 356, no. 6335 (2017): 265–70.

de Waal, Frans. "Anthropomorphism and Anthropodenial: Consistency in Our Thinking About Humans and Other Animals." *Philosophical Topics* 27, no. 1 (1999): 255–80.

Dick, Alexander J. "On the Financial Crisis, 1825–26." BRANCH: Britain, Representation, and Nineteenth-Century History, December 2012. https://branchcollective.org/?ps_articles=alexander-j-dick-on-the-financial-crisis-1825-26.

Dickens, Charles. *Bleak House*. Edited by Stephen Gill. Oxford: Oxford University Press, 2008.

———. *A Christmas Carol and Other Christmas Books*. Edited by Robert Douglas-Fairhurst. New York: Oxford University Press, 2006.

———. *Hard Times*. Edited by Paul Schlicke. Oxford: Oxford University Press, 2008.

Disraeli, Benjamin. *Sybil*. Edited by Sheila M. Smith. Oxford: Oxford University Press, 2008.

Douglass, Frederick. "Agriculture and Black Progress." In *The Frederick Douglass Papers, Series One: Speeches, Debates, and Interviews*, vol. 4, edited by John W. Blassingame and John R. McKivigan, 375–94. New Haven, CT: Yale University Press, 1991.

———. *Narrative of the Life of Frederick Douglass, an American Slave*. Edited by Deborah E. McDowell. New York: Oxford University Press, 1999.

Duncan, Ian. *Human Forms: The Novel in the Age of Evolution*. Princeton, NJ: Princeton University Press, 2019.

Dunlap, Thomas R. *Saving America's Wildlife*. Princeton, NJ: Princeton University Press, 1988.

Ebbatson, Roger. *An Imaginary England: Nation, Landscape and Literature, 1840–1920*. New York: Routledge, 2017.

Eliot, George. *Adam Bede*. Edited by Carol A. Martin. New York: Oxford University Press, 2008.

———. *Daniel Deronda*. Edited by Hugh Osborne. New York: Modern Library, 2002.

———. *The George Eliot Letters*. 9 vols. Edited by Gordon S. Haight. New Haven, CT: Yale University Press, 1954–78.

———. *The Journals of George Eliot*. Edited by Margaret Harris and Judith Johnston. Cambridge: Cambridge University Press, 1998.

———. *Middlemarch*. Edited by Rosemary Ashton. New York: Penguin, 1994.

———. *Selected Essays, Poems and Other Writings*. Edited by A. S. Byatt and Nicholas Warren. New York: Penguin, 1990.

Elton, Charles. *Animal Ecology*. New York: Macmillan, 1927.

Emerson, Ralph Waldo. "The Poet." In *Essays: Second Series*, 3–42. Boston: Houghton, Mifflin, 1876.

Enstice, Andrew. *Thomas Hardy: Landscapes of the Mind*. New York: St. Martin's Press, 1974.

Esty, Jed. *Unseasonable Youth: Modernism, Colonialism, and the Fiction of Development*. New York: Oxford University Press, 2012.

Farnie, D. A. *The English Cotton Industry and the World Market, 1815–1896*. Oxford: Clarendon Press, 1979.

Feder, Helena. *Ecocriticism and the Idea of Culture: Biology and the Bildungsroman*. London: Routledge, 2016.

Feuerstein, Anna. "The Realism of Animal Life: The Seashore, *Adam Bede*, and George Eliot's Animal Alterity." *Victorians Institute Journal* 44 (2016): 29–55.

———. "Seeing Animals on Egdon Heath: The Democratic Impulse of Thomas Hardy's *The Return of the Native*." *19: Interdisciplinary Studies in the Long Nineteenth Century* 26 (2018). doi.org/10.16995/ntn.816.

Flader, Susan L. *Thinking Like a Mountain: Aldo Leopold and the Evolution of an Ecological Attitude toward Deer, Wolves, and Forests*. Madison: University of Wisconsin Press, 1994.

Flader, Susan L., and J. Baird Callicott, eds. *The River of the Mother of God and Other Essays by Aldo Leopold*. Madison: University of Wisconsin Press, 1991.

Fleishman, Avrom. *Fiction and the Ways of Knowing: Essays on British Novels*. Austin: University of Texas Press, 1978.

Fletcher, John, and Malcolm Bradbury. "The Introverted Novel." In *Modernism, 1890–1930*, edited by Malcolm Bradbury and John Fletcher, 394–415. Sussex: Harvester Press, 1978.

Forster, E. M. *Aspects of the Novel*. New York: Harcourt, 1955.

Foster, John Bellamy. *Marx's Ecology: Materialism and Nature*. New York: Monthly Review Press, 2000.

Fraiman, Susan. *Unbecoming Women: British Women Writers and the Novel of Development*. New York: Columbia University Press, 1993.

Freedgood, Elaine. "Banishing Panic: Harriet Martineau and the Popularization of Political Economy." *Victorian Studies* 39, no. 1 (1995): 33–53. https://www.jstor.org/stable/3829415.

Frow, John. *Character and Person*. New York: Oxford University Press, 2014.

Fryckstedt, Monica Correa. "The Early Industrial Novel: *Mary Barton* and Its Predecessors." *Bulletin of the John Rylands University Library of Manchester* 63, no. 1 (1980): 12–25.

Fugate, Lauren, and John MacNeill Miller. "Shakespeare's Starlings: Literary History and the Fictions of Invasiveness." *Environmental Humanities* 13, no. 2 (2021): 301–22.

Fuller, Jennifer D. "Seeking Wild Eyre: Victorian Attitudes Towards Landscape and the Environment in Charlotte Brontë's *Jane Eyre*." *Ecozon@* 4, no. 2 (2013): 150–65.

Gadgil, Madhav, Fikret Berkes, and Carl Folke. "Indigenous Knowledge for Biodiversity Conservation." *Ambio* 22, no. 2/3 (1993): 151–56.

Galbraith, John Kenneth. *A History of Economics: The Past as the Present*. London: Penguin, 1991.

Gallagher, Catherine. *The Body Economic: Life, Death, and Sensation in Political Economy and the Victorian Novel*. Princeton, NJ: Princeton University Press, 2006.

———. *The Industrial Reformation of English Fiction: Social Discourse and Narrative Form, 1832–1867*. Chicago: University of Chicago Press, 1985.

"Gallery of Literary Characters. No. XLII. Miss Harriet Martineau." *Fraser's Magazine for Town and Country* 8, no. 47 (1833): 576.

Galor, Oded, and Omer Moav. "Natural Selection and the Origin of Economic Growth." *Quarterly Journal of Economics* 117, no. 4 (2002): 1133–91.

Galvan, Jill. "Character." *Victorian Literature and Culture* 46, no. 3/4 (2018): 612–16.

Gang, Joshua. "Mindless Modernism." *NOVEL: A Forum on Fiction* 46, no. 1 (2013): 116–32.

Garrard, Greg. *Ecocriticism*. 2nd ed. London: Routledge, 2012.

Garrett, Peter. *The Victorian Multiplot Novel: Studies in Dialogical Form*. New Haven, CT: Yale University Press, 1980.

Gaskell, Elizabeth. *Mary Barton*. Edited by Shirley Foster. New York: Oxford University Press, 2006.

Gaston, Lise. "Natural Law and Unnatural Families in Martineau's *Illustrations of Political Economy*." *Nineteenth-Century Contexts* 38, no. 3 (2016): 195–207. https://doi.org/10.1080/08905495.2016.1160530.

Ghosh, Amitav. *The Great Derangement: Climate Change and the Unthinkable*. Chicago: University of Chicago Press, 2016.

Gimingham, C. H. *An Introduction to Heathland Ecology*. Edinburgh: Oliver and Boyd, 1975.

Gissing, George. *The Collected Letters of George Gissing.* 9 vols. Edited by Paul F. Mattheisen, Arthur C. Young, and Pierre Coustillas. Athens: Ohio University Press, 1990–96.

———. *London and the Life of Literature in Late Victorian England: The Diary of George Gissing, Novelist.* Edited by Pierre Coustillas. Lewisburg, PA: Bucknell University Press, 1978.

Gold, Barri J. *Energy, Ecocriticism, and Nineteenth-Century Fiction: Novel Ecologies.* Cham, Switzerland: Palgrave Macmillan, 2021.

Golley, Frank Benjamin. *A History of the Ecosystem Concept in Ecology: More than the Sum of the Parts.* New Haven, CT: Yale University Press, 1993.

Gordon, Scott. "Darwin and Political Economy: The Connection Reconsidered." *Journal of the History of Biology* 22, no. 3 (1989): 437–59.

Gregor, Ian. *The Great Web: The Form of Hardy's Major Fiction.* Totowa, NJ: Rowman and Littlefield, 1974.

Griffiths, Devin. *The Age of Analogy: Science and Literature Between the Darwins.* Baltimore: Johns Hopkins University Press, 2016.

Griffiths, Devin, and Deanna Kreisel, eds. *After Darwin: Literature, Theory, and Criticism in the Twenty-First Century.* Cambridge: Cambridge University Press, 2023.

Grove, Richard H. *Green Imperialism: Colonial Expansion, Tropical Island Edens, and the Origins of Environmentalism, 1600–1860.* New York: Cambridge University Press, 1995.

Hadley, Elaine, Audrey Jaffe, and Sarah Winter. "Introduction: Reclaiming the Social." In *From Political Economy to Economics through Nineteenth-Century Literature,* edited by Elaine Hadley, Audrey Jaffe, and Sarah Winter, 1–25. Cham, Switzerland: Palgrave Macmillan, 2019.

Hagen, Joel B. *An Entangled Bank: The Origins of Ecosystem Ecology.* New Brunswick, NJ: Rutgers University Press, 1992.

Haight, Gordon S. *George Eliot: A Biography.* New York: Oxford University Press, 1968.

Hamera, Judith. *Parlor Ponds: The Cultural Work of the American Home Aquarium, 1850–1870.* Ann Arbor: University of Michigan Press, 2012.

Hamon, Philippe. "Rhetorical Status of the Descriptive." Translated by Ann Levonas. *New Literary History* 8, no. 1 (1976): 1–13.

Haraway, Donna J. "A Cyborg Manifesto: Science, Technology, and Socialist-Feminism in the Late Twentieth Century." In *Simians, Cyborgs, and Women: The Reinvention of Nature,* 149–81. New York: Routledge, 1991.

Hardy, Thomas. *Jude the Obscure.* Edited by Patricia Ingham. Oxford: Oxford University Press, 2008.

———. *The Life and Work of Thomas Hardy.* Edited by Michael Millgate. London: Macmillan, 1984.

———. *The Return of the Native.* Edited by Simon Gatrell. Oxford: Oxford University Press, 2005.

———. *Tess of the d'Urbervilles.* 3rd ed. Edited by Scott Elledge. New York: Norton, 1991.

Heise, Ursula K. *Imagining Extinction: The Cultural Meanings of Endangered Species.* Chicago: University of Chicago Press, 2016.

Helsinger, Elizabeth K. *Rural Scenes and National Representation: Britain, 1815–1850.* Princeton, NJ: Princeton University Press, 1997.

Hensley, Nathan K. "Database and the Future Anterior: Reading *The Mill on the Floss* Backwards." *Genre* 50, no. 1 (2017): 117–37.

———. *Forms of Empire: The Poetics of Victorian Sovereignty.* New York: Oxford University Press, 2016.

Hensley, Nathan K., and Philip Steer, eds. *Ecological Form: System and Aesthetics in the Age of Empire.* New York: Fordham University Press, 2019.

Hildebrand, Jayne. "Environmental Desire in *The Mill on the Floss*." *Nineteenth-Century Literature* 76, no. 2 (2021): 192–222.

———. "*Middlemarch*'s Medium: Description, Sympathy, and Ambient Worlds." *ELH: English Literary History* 85, no. 4 (2018): 999–1023.

Hodgson, Geoffrey M., and Thorbjørn Knudsen. *Darwin's Conjecture: The Search for General Principles of Social and Economic Evolution.* Chicago: University of Chicago Press, 2010.

Hopkins, A. B. *Elizabeth Gaskell: Her Life and Work.* New York: Octagon Books, 1971.

Hovanec, Caroline. *Animal Subjects: Literature, Zoology, and British Modernism.* Cambridge: Cambridge University Press, 2018.

Howarth, Glennys. "Professionalizing the Funeral Industry in England, 1700–1960." In *The Changing Face of Death: Historical Accounts of Death and Disposal,* edited by Peter C. Jupp and Glennys Howarth, 120–34. London: Palgrave Macmillan, 1997.

Howe, Irving. "The Return of the Native." In *Thomas Hardy's "Return of the Native,"* edited by Harold Bloom, 13–20. New York: Chelsea House, 1987.

Howkins, Chris. *Heathers and Heathlands.* Addlestone, UK: Chris Howkins, 2004.

Humphrey, Robert. *Stream of Consciousness in the Modern Novel.* Berkeley: University of California Press, 1972.

Huzzey, Richard. *Freedom Burning: Anti-Slavery and Empire in Victorian Britain.* Ithaca, NY: Cornell University Press, 2012.

"*Illustrations of Political Economy.* Nos. 1–12. By Harriet Martineau. London, 1832–1833." *Quarterly Review* 49 (April–July 1833): 136–52. https://babel.hathitrust.org/cgi/pt?id=uc1.a0010589935&seq=148.

Irwin, Michael. *Reading Hardy's Landscapes.* New York: St. Martin's Press, 2000.

Jackson, Zakiyyah Iman. *Becoming Human: Matter and Meaning in an Antiblack World.* New York: New York University Press, 2020.

Jacobus, Mary. "The Question of Language: Men of Maxims and *The Mill on the Floss*." *Critical Inquiry* 8, no. 2 (1981): 207–22.

Jakobson, Roman, and Morris Halle. *Fundamentals of Language.* 2nd ed. Berlin: Mouton de Gruyter, 2002.

James, Henry. "Preface to *Roderick Hudson*." In *Literary Criticism: French Writers, Other European Writers, The Prefaces to the New York Edition,* edited by Leon Edel, 1039–52. New York: Library of America, 1984.

Jameson, Fredric. Introduction to *The Historical Novel*, by George Lukács, 1–8. Translated by Hannah and Stanley Mitchell. Lincoln: University of Nebraska Press, 1983.

Jardine, N., J. A. Secord, and E. C. Spary, eds. *Cultures of Natural History*. Cambridge: Cambridge University Press, 1996.

Jay, Martin. *Marxism and Totality: The Adventures of a Concept from Lukács to Habermas*. Berkeley: University of California Press, 1984.

———. "Modernism and the Specter of Psychologism." *Modernism/Modernity* 3, no. 2 (1996): 93–111.

Jevons, W. Stanley. *The Theory of Political Economy*. 4th ed. London: Macmillan, 1924.

Jonsson, Fredrik Albritton. *Enlightenment's Frontier: The Scottish Highlands and the Origins of Environmentalism*. New Haven, CT: Yale University Press, 2013.

———. "Island, Nation, Planet: Malthus in the Enlightenment." In *New Perspectives on Malthus*, edited by Robert J. Mayhew, 128–52. Cambridge: Cambridge University Press, 2016.

Kahneman, Daniel, and Amos Tversky. "Prospect Theory: An Analysis of Decision Under Risk." *Econometrica* 47, no. 2 (1979): 263–92.

Kapp, K. William. "Environmental Disruption and Social Costs: A Challenge to Economics." In *Political Economy of Environment: Problems of Method*, 91–102. Paris: École Pratique des Hautes Études and Mouton, 1972.

Kennedy, Margaret S. "A Breath of Fresh Air: Eco-Consciousness in *Mary Barton* and *Jane Eyre*." *Victorian Literature and Culture* 45, no. 3 (2017): 509–26.

Keogh, Luke. *The Wardian Case: How a Simple Box Moved Plants and Changed the World*. Chicago: University of Chicago Press, 2020.

Kerridge, Richard. "Ecological Hardy." In *Beyond Nature Writing: Expanding the Boundaries of Ecocriticism*, edited by Karla Armbruster and Kathleen R. Wallace, 126–42. Charlottesville: University of Virginia Press, 2001.

Kettle, Arnold. "The Early Victorian Social-Problem Novel." In *The Pelican Guide to English Literature*, vol. 6, edited by Boris Ford, 169–87. New York: Penguin, 1958.

Kim, Claire Jean. *Dangerous Crossings: Race, Species, and Nature in a Multicultural Age*. New York: Cambridge University Press, 2015.

Kimmerer, Robin Wall. *Braiding Sweetgrass: Indigenous Wisdom, Scientific Knowledge, and the Teachings of Plants*. Minneapolis: Milkweed Editions, 2013.

King, Amy M. *The Divine in the Commonplace: Reverent Natural History and the Novel in Britain*. New York: Cambridge University Press, 2019.

Kingsley, Charles. *Alton Locke, Tailor and Poet*. London: Cassell, 1967.

Kingstone, Helen. "Human-Animal Elision: A Darwinian Universe in George Eliot's Novels." *Nineteenth-Century Contexts* 40, no. 1 (2018): 87–103.

Klancher, Jon. *Transfiguring the Arts and Sciences: Knowledge and Cultural Institutions in the Romantic Age*. New York: Cambridge University Press, 2013.

Klaver, Claudia C. *A/Moral Economics: Classical Political Economy and Cultural Authority in Nineteenth-Century England*. Columbus: Ohio State University Press, 2003.

———. "Imperial Economics: Harriet Martineau's *Illustrations of Political Economy* and the Narration of Empire." *Victorian Literature and Culture* 35, no. 1 (2007): 21–40. http://www.jstor.org/stable/40347122.

Klaver, J. M. I. *The Apostle of the Flesh: A Critical Life of Charles Kingsley*. Leiden: Brill, 2006.

Krasner, James. *The Entangled Eye: Visual Perception and the Representation of Nature in Post-Darwinian Narrative*. Oxford: Oxford University Press, 1992.

Kreilkamp, Ivan. *Minor Creatures: Persons, Animals, and the Victorian Novel*. Chicago: University of Chicago Press, 2018.

Lakoff, George, and Mark Johnson. *Metaphors We Live By*. 2nd ed. Chicago: University of Chicago Press, 2003.

Lamb, John B. "Storied Matter: Waste and Wastelands in Thomas Hardy's *Return of the Native*." *ISLE: Interdisciplinary Studies in Literature and Environment* 30, no. 4 (2021): 972–97. doi.org/10.1093/isle/isab077.

Lane, Christopher. *The Age of Doubt: Tracing the Roots of Our Religious Uncertainty*. New Haven, CT: Yale University Press, 2011.

Langbaum, Robert. *Thomas Hardy in Our Time*. New York: St. Martin's Press, 1995.

Latour, Bruno. *Facing Gaia: Eight Lectures on the New Climatic Regime*. Translated by Catherine Porter. Cambridge: Polity Press, 2017.

———. *Politics of Nature: How to Bring the Sciences into Democracy*. Translated by Catherine Porter. Cambridge, MA: Harvard University Press, 2004.

Law, Jules. *The Social Life of Fluids: Blood, Milk, and Water in the Victorian Novel*. Ithaca, NY: Cornell University Press, 2010.

Lea, Hermann. *Thomas Hardy's Wessex*. London: Macmillan, 1913.

Lear, Linda. *Rachel Carson: Witness for Nature*. New York: Henry Holt, 1997.

Leopold, Aldo. *A Sand County Almanac with Essays on Conservation from Round River*. New York: Ballantine Books, 1970.

Lesjak, Carolyn. *The Afterlife of Enclosure: British Realism, Character, and the Commons*. Stanford, CA: Stanford University Press, 2021.

Levine, Caroline. *Forms: Whole, Rhythm, Hierarchy, Network*. Princeton, NJ: Princeton University Press, 2015.

———. "Narrative Networks: *Bleak House* and the Affordances of Form." *NOVEL: A Forum on Fiction* 42, no. 3 (2009): 517–23.

Levine, George. *Darwin the Writer*. Oxford: Oxford University Press, 2011.

———, ed. *One Culture: Essays in Science and Literature*. Madison: University of Wisconsin Press, 1987.

———. *Reading Thomas Hardy*. Cambridge: Cambridge University Press, 2017.

———. *Realism, Ethics and Secularism: Essays on Victorian Literature and Science*. New York: Cambridge University Press, 2008.

———. *The Realistic Imagination: English Fiction from Frankenstein to Lady Chatterley*. Chicago: University of Chicago Press, 1981.

Lovejoy, Arthur O. *The Great Chain of Being: A Study of the History of an Idea*. Cambridge, MA: Harvard University Press, 1961.

Lutts, Ralph H. "Animals and Ideas." In *The Wild Animal Story*, edited by Ralph H. Lutts, 1–21. Philadelphia: Temple University Press, 1998.

Lytle, Mark Hamilton. *The Gentle Subversive: Rachel Carson, Silent Spring, and the Rise of the Environmental Movement*. New York: Oxford University Press, 2007.

MacDorman, Karl F. "Androids as an Experimental Apparatus: Why Is There an Uncanny Valley and Can We Exploit It?" *Proceedings of the CogSci 2005 Workshop: Toward Social Mechanisms of Android Science*, January 2005, 106–18. www.researchgate.net/publication/245406914_Androids_as_an_Experimental_Apparatus_Why_Is_There_an_Uncanny_Valley_and_Can_We_Exploit_It.

MacDuffie, Allen. *Victorian Literature, Energy, and the Ecological Imagination*. New York: Cambridge University Press, 2014.

Macfarlane, Robert. *Landmarks*. London: Penguin Random House UK, 2016.

Malthus, Thomas. *An Essay on the Principle of Population*. Edited by Geoffrey Gilbert. Oxford: Oxford University Press, 2008.

———. *Principles of Political Economy*. 2 vols. Edited by John Pullen. Cambridge: Cambridge University Press, 1989.

Mann, Abigail, and Kathleen Bérés Rogers. "Objects and Objectivity: Harriet Martineau as Nineteenth-Century Cyborg." *Prose Studies* 33, no. 3 (2011): 241–56. doi.org/10.1080/01440357.2011.647271.

Martineau, Harriet. *Autobiography and Memorials of Harriet Martineau*. 2 vols. Edited by Maria Weston Chapman. Boston: James R. Osgood, 1877.

———. "Cinnamon and Pearls." In vol. 7 of *Illustrations of Political Economy*. London: Charles Fox, 1833. Online Library of Liberty. https://oll.libertyfund.org/title/martineau-illustrations-of-political-economy-vol-7.

———. *Illustrations of Political Economy: Selected Tales*. Edited by Deborah Anna Logan. Peterborough: Broadview, 2004.

Mayhew, Robert J. *Malthus: The Life and Legacies of an Untimely Prophet*. Cambridge, MA: Harvard University Press, 2014.

———, ed. *New Perspectives on Malthus*. Cambridge: Cambridge University Press, 2016.

McAuley, Kyle. "George Eliot's Estuarial Form." *Victorian Literature and Culture* 48, no. 1 (2020): 187–217.

McCarthy, Jeffrey Mathes. *Green Modernism: Nature and the English Novel, 1900 to 1930*. New York: Palgrave Macmillan, 2015.

McCloskey, Deirdre N. *The Rhetoric of Economics*. 2nd ed. Madison: University of Wisconsin Press, 1998.

McIntosh, Robert P. *The Background of Ecology: Concept and Theory*. New York: Cambridge University Press, 1985.

McLane, Maureen N. "Malthus Our Contemporary? Toward a Political Economy of Sex." *Studies in Romanticism* 52, no. 3 (2013): 337–62. www.jstor.org/stable/24247336.

McLellan, David, ed. *Karl Marx: Selected Writings*. 2nd ed. New York: Oxford University Press, 2000.

Mead, Rebecca. *My Life in Middlemarch*. New York: Crown, 2014.

Menely, Tobias. *Climate and the Making of Worlds: Toward a Geohistorical Poetics*. Chicago: University of Chicago Press, 2021.

Menely, Tobias, and Jesse Oak Taylor, eds. *Anthropocene Reading: Literary History in Geologic Times*. University Park: Pennsylvania State University Press, 2017.

Mill, James. *Elements of Political Economy.* 3rd ed. London: Henry G. Bohn, 1844.
Miller, Elizabeth Carolyn. "Dendrography and Ecological Realism." *Victorian Studies* 58, no. 4 (2016): 698–718.
———. "Ecology." *Victorian Literature and Culture* 46, no. 3/4 (2018): 653–56.
———. *Extraction Ecologies and the Literature of the Long Exhaustion.* Princeton, NJ: Princeton University Press, 2021.
Miller, J. Hillis. *Thomas Hardy: Distance and Desire.* Cambridge, MA: Belknap Press, 1970.
Miller, John MacNeill. "Weird Beyond Description: Weird Fiction and the Suspicion of Scenery." *Victorian Studies* 62, no. 2 (2020): 244–52.
Miller, Robyn. "'Resolute, Wild, and Free': Women's Leisure and Avian Ecologies in *Jane Eyre.*" *Nineteenth-Century Gender Studies* 15, no. 2 (2019). http://ww.w.ncgsjournal.com/issue152/miller.html.
Millgate, Michael. *Thomas Hardy: A Biography Revisited.* Oxford: Oxford University Press, 2004.
Moretti, Franco. *The Ways of the World: The* Bildungsroman *in European Culture.* New ed. Translated by Albert Spragia. London: Verso, 2000.
Morgan, Kenneth. *Slavery and the British Empire: From Africa to America.* New York: Oxford University Press, 2007.
Morton, Timothy. *Dark Ecology: For a Logic of Future Coexistence.* New York: Columbia University Press, 2016.
———. *The Ecological Thought.* Cambridge, MA: Harvard University Press, 2010.
———. *Ecology Without Nature: Rethinking Environmental Aesthetics.* Cambridge, MA: Harvard University Press, 2007.
Myers, Greg. *Writing Biology: Texts in the Social Construction of Scientific Knowledge.* Madison: University of Wisconsin Press, 1990.
Nash, Roderick. *Wilderness and the American Mind.* 3rd ed. New Haven, CT: Yale University Press, 1982.
Nelson, Melissa K., and Daniel Shilling, eds. *Traditional Ecological Knowledge: Learning from Indigenous Practices for Environmental Sustainability.* Cambridge: Cambridge University Press, 2018.
Nelson, Richard R., and Sidney G. Winter. *An Evolutionary Theory of Economic Change.* Cambridge, MA: Belknap Press, 1982.
Ngai, Sianne. *Our Aesthetic Categories: Zany, Cute, Interesting.* Cambridge, MA: Harvard University Press, 2012.
Nord, Deborah Epstein. *Gypsies and the British Imagination, 1807–1930.* New York: Columbia University Press, 2006.
Oakes, Jill, Rick Riewe, Kimberley Wilde, Alison Edmunds, and Alison Dubois, eds. *Native Voices in Research.* Winnipeg: Aboriginal Issues Press, 2003.
O'Gorman, Francis. "The Rural Scene: Victorian Literature and the Natural World." In *The Cambridge History of Victorian Literature,* edited by Kate Flint, 532–49. Cambridge: Cambridge University Press, 2012.
Olson, Richard G. *Science and Religion, 1450–1900.* Westport, CT: Greenwood Press, 2004.

Padian, Kevin. "Evolution and Deep Time in Selected Works of Hardy." In *The Ashgate Research Companion to Thomas Hardy*, edited by Rosemarie Morgan, 217–33. Farnham, UK: Ashgate, 2010.

Paradis, James. "Darwin and Landscape." *Annals of the New York Academy of Sciences* 360, no. 1 (1981): 85–110.

Parry, James. *Living Landscapes: Heathland*. London: The National Trust, 2003.

Pawelec, Andrzej. "The Death of Metaphor." *Studia Linguistica* 123 (2006): 117–21.

Pearce, Trevor. "From 'Circumstances' to 'Environment': Herbert Spencer and the Origins of the Idea of Organism-Environment Interaction." *Studies in the History and Philosophy of Biological and Biomedical Sciences* 41 (2010): 241–52.

Peterson, Linda H. "From French Revolution to English Reform: Hannah More, Harriet Martineau, and the 'Little Book.'" *Nineteenth-Century Literature* 60, no. 4 (2006): 409–50. https://www.jstor.org/stable/10.1525/ncl.2006.60.4.409.

Pettitt, Clare. *Serial Revolutions 1848: Writing, Politics, Form*. Oxford: Oxford University Press, 2022.

Pielak, Chase. "Hunting Gwendolen: Animetaphor in *Daniel Deronda*." *Victorian Literature and Culture* 40, no. 1 (2012): 99–115.

Pizzo, Justine. "Atmospheric Exceptionalism in *Jane Eyre*: Charlotte Brontë's Weather Wisdom." *PMLA* 131, no. 1 (2016): 84–100.

Poovey, Mary. *Making a Social Body: British Cultural Formation, 1830–1864*. Chicago: University of Chicago Press, 1995.

Porter, Theodore M. *Trust in Numbers: The Pursuit of Objectivity in Science and Public Life*. Princeton, NJ: Princeton University Press, 1995.

Preston, Christopher J. *Tenacious Beasts: Wildlife Recoveries That Change How We Think about Animals*. Cambridge, MA: The MIT Press, 2023.

Qualls, Barry. "George Eliot and Religion." In *The Cambridge Companion to George Eliot*, edited by George Levine, 119–37. Cambridge: Cambridge University Press, 2001.

Rackham, Oliver. *The History of the Countryside: The Classic History of Britain's Landscape, Flora and Fauna*. London: Phoenix, 2004.

Richardson, Rebecca. "Environmental and Economic Systems in Harriet Martineau's *Illustrations of Political Economy*." *Nineteenth-Century Prose* 47, no. 2 (2020): 61–88.

Ridley, Matt. "The Natural Order of Things." *Spectator*, January 7, 2009. https://www.spectator.co.uk/2009/01/the-natural-order-of-things.

Ritvo, Harriet. *The Dawn of Green: Manchester, Thirlmere, and Modern Environmentalism*. Chicago: University of Chicago Press, 2009.

———. *The Platypus and the Mermaid, and Other Figments of the Classifying Imagination*. Cambridge, MA: Harvard University Press, 1997.

Robbins, Lionel. *A History of Economic Thought: The LSE Lectures*. Edited by Steven G. Medema and Warren J. Samuels. Princeton, NJ: Princeton University Press, 1998.

Robertson, Priscilla. *Revolutions of 1848: A Social History*. Princeton, NJ: Princeton University Press, 1952.

Rome, Adam. *The Bulldozer in the Countryside: Suburban Sprawl and the Rise of American Environmentalism*. Cambridge: Cambridge University Press, 2001.

———. *The Genius of Earth Day: How a 1970 Teach-In Unexpectedly Made the First Green Generation*. New York: Hill and Wang, 2013.

Ross, Shawna. *Charlotte Brontë at the Anthropocene*. Albany: State University of New York Press, 2020.

Ross, Sydney. "*Scientist:* The Story of a Word." *Annals of Science* 18, no. 2 (1962): 65–85. https://www.tandfonline.com/doi/pdf/10.1080/00033796200202722.

Sanders, Mike. "From 'Political' to 'Human' Economy: The Visions of Harriet Martineau and Frances Wright." *Women: A Cultural Review* 12, no. 2 (2001): 192–203. https://doi.org/10.1080/095740400110060238.

Schneider-Mayerson, Matthew, Alexa Weik von Mossner, W. P. Malecki, and Frank Hakemulder, eds. *Empirical Ecocriticism: Environmental Narratives for Social Change*. Minneapolis: University of Minnesota Press, 2023.

Schumpeter, Joseph A. *History of Economic Analysis*. Edited by Elizabeth Boody Schumpeter. New York: Oxford University Press, 1954.

Schuster, Joshua. *The Ecology of Modernism: American Environments and Avant-Garde Poetics*. Tuscaloosa: University of Alabama Press, 2015.

Scott, Heidi C. M. *Chaos and Cosmos: Literary Roots of Modern Ecology in the British Nineteenth Century*. University Park: Pennsylvania State University Press, 2014.

Secord, Anne. "Corresponding Interests: Artisans and Gentlemen in Nineteenth-Century Natural History." *British Journal for the History of Science* 27, no. 4 (1994): 383–408.

———. "Elizabeth Gaskell's Social Vision: The Natural Histories of *Mary Barton*." In *Uncommon Contexts: Encounters Between Science and Literature, 1800–1914*, edited by Ben Marsden, Hazel Hutchison, and Ralph O'Connor, 125–44. Pittsburgh: University of Pittsburgh Press, 2016.

———. "Science in the Pub: Artisan Botanists in Early Nineteenth-Century Lancashire." *History of Science* 32, no. 3 (1994): 269–315.

Sellers, Christopher C. *Crabgrass Crucible: Suburban Nature and the Rise of Environmentalism in Twentieth-Century America*. Chapel Hill: University of North Carolina Press, 2012.

Serpell, C. Namwali. "Mutual Exclusion, Oscillation, and Ethical Projection in *The Crying of Lot 49* and *The Turn of the Screw*." *Narrative* 16, no. 3 (2008): 223–55.

Shuttleworth, Sally. *George Eliot and Nineteenth-Century Science: The Make-Believe of a Beginning*. Cambridge: Cambridge University Press, 1984.

Simmons, James Richard, Jr. "Industrial and 'Condition of England' Novels." In *A Companion to the Victorian Novel*, edited by Patrick Brantlinger and William B. Thesing, 336–52. Oxford: Blackwell, 2002.

Slotten, Ross A. *The Heretic in Darwin's Court: The Life of Alfred Russel Wallace*. New York: Columbia University Press, 2004.

Smith, R. E. "George Henry Lewes and His 'Physiology of Common Life,' 1859." *Proceedings of the Royal Society of Medicine* 53 (July 1960): 569–74. https://journals.sagepub.com/doi/pdf/10.1177/003591576005300722.

Southwood, Richard, and J. R. Clarke. "Charles Sutherland Elton, 29 March 1900–1 May 1991." *Biographical Memories of Fellows of the Royal Society London* 45 (1999): 129–46.

Spencer, Herbert. "The Social Organism." *Westminster and Foreign Quarterly Review*, January 1860. https://babel.hathitrust.org/cgi/pt?id=mdp.39015009231849.

Sultzbach, Kelly. *Ecocriticism in the Modernist Imagination: Forster, Woolf, and Auden*. New York: Cambridge University Press, 2016.

Swanson, Heather, Anna Tsing, Nils Bubandt, and Elaine Gan. "Bodies Tumbled into Bodies." In *Arts of Living on a Damaged Planet*, edited by Anna Tsing, Heather Swanson, Elaine Gan, and Nils Bubandt, M1–M14. Minneapolis: University of Minnesota Press, 2017.

Taylor, Jesse Oak. *The Sky of Our Manufacture: The London Fog in British Fiction from Dickens to Woolf*. Charlottesville: University of Virginia Press, 2016.

Thomas, Hugh. *The Slave Trade: The History of the Atlantic Slave Trade, 1440–1870*. New York: Picador, 1997.

Tilley, Christopher. *Landscape in the Long Durée: A History and Theory of Pebbles in a Pebbled Heathland Landscape*. London: University College London Press, 2017.

Trollope, Anthony. *An Autobiography*. New York: Dodd, Mead, 1927.

Tsing, Anna. "Arts of Inclusion, or How to Love a Mushroom." *Manoa* 22, no. 2 (2010): 191–203.

Tyndall, Gillian. *The Born Exile: George Gissing*. New York: Harcourt Brace Jovanovich, 1974.

Uglow, Jenny. *Elizabeth Gaskell: A Habit of Stories*. New York: Farrar, Straus and Giroux, 1993.

Vermeij, Geerat J. *Nature: An Economic History*. Princeton, NJ: Princeton University Press, 2004.

Wallace, Alfred Russel. *The Annotated Malay Archipelago*. Edited by John van Whye. Singapore: National University of Singapore Press, 2017.

Walvin, James. *England, Slaves, and Freedom, 1776–1838*. Jackson: University Press of Mississippi, 1986.

———. *Making the Black Atlantic: Britain and the African Diaspora*. London: Cassell, 2000.

Warren, Julianne Lutz. *Aldo Leopold's Odyssey*. 10th anniversary ed. Washington, DC: Island Press, 2016.

Watts, May Theilgaard. *Reading the Landscape of America*. Rochester, NY: Nature Study Guild Publishers, 1975.

Webb, R. K. *Harriet Martineau: A Radical Victorian*. New York: Columbia University Press, 1960.

Wessels, Tom. *Reading the Forested Landscape: A Natural History of New England*. Illustrated by Brian D. Cohen. Woodstock, VT: Countryman Press, 1997.

Wilk, Elvia. *Death by Landscape*. New York: Soft Skull, 2022.

Williams, Merryn. *Thomas Hardy and Rural England*. New York: Columbia University Press, 1972.

Williams, Raymond. *The Country and the City*. New York: Oxford University Press, 1975.

———. *Culture and Society: 1780–1950*. 2nd ed. New York: Columbia University Press, 1983.

———. "Realism." In *Keywords: A Vocabulary of Culture and Society*, 198–202. New ed. New York: Oxford University Press, 2015.

Wilson, E. O. *Sociobiology: The New Synthesis*. 25th anniversary ed. Cambridge, MA: Harvard University Press, 2000.

Winter, James. *Secure from Rash Assault: Sustaining the Victorian Environment*. Berkeley: University of California Press, 1999.

Woloch, Alex. *The One vs. the Many: Minor Characters and the Space of the Protagonist in the Novel*. Princeton, NJ: Princeton University Press, 2003.

Worster, Donald. *Nature's Economy: A History of Ecological Ideas*. 2nd ed. New York: Cambridge University Press, 1994.

Wu, Kaidu, and David Dunning. "Hypocognition: Making Sense of the Landscape Beyond One's Conceptual Reach." *Review of General Psychology* 22, no. 1 (2018): 25–35.

Wynter, Sylvia. "Unsettling the Coloniality of Being/Power/Truth/Freedom: Towards the Human, After Man, Its Overrepresentation—An Argument." *CR: The New Centennial Review* 3, no. 3 (2003): 257–337.

Wynter, Sylvia, and Katherine McKittrick. "Unparalleled Catastrophe for Our Species? Or, to Give Humanness a Different Future: Conversations." In *Sylvia Wynter: On Being Human as Praxis*, edited by Katherine McKittrick, 9–89. Durham, NC: Duke University Press, 2015.

Yeo, Richard. *Defining Science: William Whewell, Natural Knowledge, and Public Debate in Early Victorian Britain*. Cambridge: Cambridge University Press, 1993.

Index

abolitionism, 46, 59–61
Adam Bede (Eliot), 14, 78, 79, 86–95, 175n32
Adams, James Eli, 175n32
affordance theory, 2, 40, 163n3, 169n64
Alaimo, Stacy, 26–27
Allen, David Elliston, 70
Alton Locke (Kingsley), 54, 58–59, 60–61, 64, 74, 75, 173n60
Animal Ecology (Elton), 141–46, 147, 149
animals. *See* bees; birds; cattle; dogs; horses; insects; livestock; marine life; orangutans; worms
animal studies, 11, 172n37, 183n38
Anthropocene, 173n58
anthropocentrism, 14, 23–24, 29, 76, 80, 98, 108–12, 115, 153, 167n26, 171n28
anthropomorphism, 56, 72, 90–91, 118, 121, 125, 127, 130, 158, 179n31. *See also* characterization; personification
apocalypse narratives, 160–61, 183n40
Austen, Jane, 46
Autobiography (Darwin), 37
Autobiography (Martineau), 30–31

background, as literary device, 13, 14, 32, 35–37, 47, 49, 89, 100–108, 110–11, 117, 136, 138, 162. *See also* ecological literacy; landscape; setting
Baldwin, Rosecrans, 158
Bambi (Salten), 158
Banerjee, Sukanya, 170n10
Barker, Roy, 152
Beagle, HMS, 17, 37, 66

Beer, Gillian, 9, 125, 132
bees, 39, 169n62. *See also* insects
Bennett, Joshua, 171n28, 172n37
Bentham, Jeremy, 43
bildungsroman, 95, 97–100, 175n39
birds, 17, 36, 39, 64, 70, 93, 96, 128, 137, 143, 145, 147, 151, 152–55
Blackwood's Edinburgh Magazine, 78
Bleak House (Dickens), 54–55
Boisseron, Bénédicte, 11, 165n13, 172n37
boredom, 101–5, 110, 117, 120–21, 129
Brontë, Charlotte, 4, 43

Carlyle, Jane Welsh, 86–87
Carlyle, Thomas, 43, 86–87
Carr-Saunders, Alexander, 143
Carson, Rachel, 149, 151–55, 181nn19–20, 182n32
cattle, 23–25, 28, 36, 39, 59, 60, 63, 64, 131, 133, 144, 145, 152. *See also* horses; livestock
Çelikkol, Ayşe, 167n34
Chapman, John, 80–81
Chapman, Maria Weston, 46, 170n7
characterization, 14, 34, 49, 79–80, 88–89, 99–100, 102–4, 106–8, 110, 117, 128, 135–36, 138, 151, 155, 159–61, 174n28, 177n65; of animals, 56, 64, 74, 79–80, 86–95, 104–5, 111, 131, 135, 156, 158, 175n29, 175n32, 176n59, 182n37; of landscapes, 116, 118, 120–26, 178n12; of nature, 116, 126–27, 132, 137; of plants, 130–31, 135. *See also* anthropomorphism; personification

Chartism, 42–43, 47, 53, 68
Chaudhuri, Una, 183n38
Chez, Keridiana W., 175n30
Choi, Tina Young, 9, 165n12
"Cinnamon and Pearls" (Martineau), 18, 19, 32–33, 63
Cohen, Jeffrey Jerome, 135
Coleridge, Samuel Taylor, 101, 146
coming-of-age story. *See* bildungsroman
commodification, 45, 56–61, 65–66, 68–69, 73, 86, 160. *See also* dehumanization
Commoner, Barry, 3
Communist Manifesto, The (Marx), 42
Comte, Auguste, 105
Condition of England novels. *See* industrial novels
Condition of England Question, 43, 76
Condorcet, Marquis de, 20, 21, 22
conservation movements, 32–33, 117, 127, 145, 146–47. *See also* environmentalism; values: ecology and
Corn Laws, 42
Courtemanche, Eleanor, 46
crops, 23–25, 28, 32, 33–34, 36, 42, 56, 58, 62, 68, 103, 104, 111

Daniel Deronda (Eliot), 108, 110–11
Darwin, Caroline, 37
Darwin, Charles, 7, 13, 144; *Autobiography*, 37; biography of, 17–18, 37–38; ecological plotting of, 38–40, 105, 112, 115, 129–32, 135–36, 139, 141, 152; Eliot on, 112, 177n66; *The Formation of Vegetable Mould through the Action of Worms*, 132–33; on heathlands, 14, 39, 115, 128–36, 141; influence of, 13, 76, 112, 125, 135, 139, 143, 175n36; Malthus and, 38, 40, 44, 66, 70, 115, 165n11, 168nn57–58; Martineau and, 18, 19, 37–38, 44; *On the Origin of Species*, 14, 20, 29, 38–39, 40, 44, 112, 115, 128, 133; theory of natural selection by, 20, 38, 70–71, 132
Darwin, Erasmus Alvey, 19, 38

Darwin, Susan, 17–19, 32, 37
DDT poisoning, 152–53, 181n19. *See also* pesticides
Death of Ivan Ilyich, The (Tolstoy), 158
deep time, 121, 132, 135, 180n38
dehumanization, 56–61, 64–65, 96, 124. *See also* commodification
"Demerara" (Martineau), 63
description, as literary device, 101–4, 108, 110–11, 115, 117, 150, 168n45, 176n47
Dickens, Charles, 8, 43; *Bleak House*, 54–55; *Hard Times*, 64, 69–70, 74
Dimock, Wai Chee, 183n45
disciplines, academic: benefits of specialization in, 15; history of, 163n1; linkages between, 1, 5, 7–8, 9, 18, 20, 159–61; separation of, 5, 6, 7, 8, 9, 13–15, 20, 44–45, 51–52, 55, 114, 116, 148, 159; shortcomings of specialization in, 15, 116, 138–40, 159–62. *See also* interdisciplinarity
Disraeli, Benjamin, 43; *Sybil*, 54, 58, 64, 74
dogs, 74, 87–93, 96, 104–7, 108, 111, 143, 158–59, 160, 162, 175n30
Domesday Book, 122
Douglass, Frederick, 59–60, 171n28
Doyle, Arthur Conan, 157

ecocriticism, 6–8, 157, 164nn8–9, 175n31, 177n4, 182n35, 183n45
ecological literacy, 14, 35–37, 132–33, 155
ecological plot, 3–6, 7, 8, 9, 11, 14, 44, 47, 59, 62–64, 71–72, 80, 91, 108, 120, 128, 139–40, 153–55, 157, 183n45. *See also* plot, as literary device
ecological thought, the, 26, 164n4, 166n17
ecology, 3, 7, 11, 13, 26–27, 128, 132, 141–42, 144, 145, 157, 160; economics and, 7–8, 18–20, 28–29, 55–56, 138–40, 143, 145–50, 159–61, 183nn41–42; literary history of, overview, 1–16, 142; as offshoot of natural history, 5, 15, 18–20, 38–40, 44–45, 51, 71, 112, 138–40,

142–43; political, 28; social novel and, 5, 7, 114–16, 126–30, 133–36, 138–40, 152–53, 159. *See also* environmentalism; Oekologie
ecomimesis, 104
economics, 5, 7, 9, 11, 12, 13, 22, 154, 160–61; ecology and, 7–8, 18–20, 28–29, 55–56, 138–40, 143, 145–50, 159–61, 183nn41–42; as offshoot of political economy, 5, 9, 15, 18–20, 21, 44–45, 51, 82–84, 138–40, 174n14; social novel and, 5, 7–8, 18–20, 116, 138–40, 159; as term, 7. *See also* political economy
Edgeworth, Maria, 46, 47
elegy, environmental, 137, 160, 183n40
Elements of Political Economy (Mill), 83
Eliot, George, 7, 8, 46, 78–79, 128, 144; *Adam Bede*, 14, 78, 79, 86–95, 175n32; biography of, 78–79, 80–82; *Daniel Deronda*, 108, 110–11; Darwin and, 112, 175n36, 176n59, 177n66; ecological plotting of, 91–92, 111–12; *Felix Holt, the Radical*, 170n6; Hardy and, 115–16, 178n8; influence of, 80, 81; on Martineau, 46; *Middlemarch*, 80, 92, 108–9, 158; *The Mill on the Floss*, 14, 79–80, 91–92, 95–108, 110, 112; "The Natural History of German Life," 13–14, 81–82, 84–85, 94; reception of, 81, 95, 109; religion and, 81, 86, 88, 89–90, 95, 100, 110–11
Elton, Charles, 141–46, 147, 149, 181n6
Elton, Letitia MacColl, 141
Elton, Oliver, 141
Emerson, Ralph Waldo, 6, 119
Enlightenment, the, 22–24, 33, 59–60, 63–64, 168n43
environment, as term, 105, 117
environmentalism, 6, 15, 18, 21, 29, 37, 114, 116, 117, 127, 136, 137, 149, 160, 164n5, 178n11, 183n40. *See also* conservation movements; values, ecology and
Essay on the Principle of Population (Malthus), 12, 19–29, 32, 62, 69, 82, 166n17, 167n26; second edition of, 25, 33, 82, 166n20, 167n26
Essence of Christianity, The (Feuerbach), 81
ethics. *See* values
Europe: influence on Britain of, 12, 20, 42, 76, 81–82, 84, 85, 105, 141; revolutions in, 20, 42, 45
Evans, Marian. *See* Eliot, George

Feder, Helena, 98
Felix Holt, the Radical (Eliot), 170n6
Feuerbach, Ludwig, 76, 81
fish. *See* marine life
fisheries, 32, 33–34, 62, 65, 145. *See also* marine life
food chains, 3, 21, 24, 29, 115, 143, 144, 151, 154
forestry, 3, 36, 71, 91, 116, 131, 136, 137, 138, 146, 147, 148
Formation of Vegetable Mould through the Action of Worms, The (Darwin), 132–33
Forster, E. M., 102
Frankenstein (Shelley), 52

Gallagher, Catherine, 26
Garrett, Peter, 9
Gaskell, Elizabeth, 7, 43; biography of, 46, 47, 55; ecological plotting of, 13, 47–51, 152; Martineau and, 46, 47, 51, 55; *Mary Barton*, 13, 45, 46–54, 55, 56–58, 64–66, 66–69, 70, 73, 74–75; religion and, 46, 53, 74–75
genre fiction, 157, 182nn36–37
Ghosh, Amitav, 177n65
Gissing, George, 113–14
Godwin, William, 12, 20, 21, 22
Gosse, Philip Henry, 70
Gradgrind, as term, 69–70
Griffiths, Devin, 132, 168n54
Grove, Richard, 32–33

Haeckel, Ernst, 39, 112, 128, 139, 141
Hard Times (Dickens), 64, 69–70, 74

Hardy, Thomas, 7, 112, 157; biography of, 113–15; Darwin and, 115–16, 125, 132, 135–36, 177n6; Eliot and, 115–16, 178n8; Gissing and, 113–14; on heathlands, 115–16, 117–27, 133–36, 137; influence of, 137; *Jude the Obscure*, 126; reception of, 113–14, 127, 132, 183n45; *The Return of the Native*, 14, 115, 116–27, 133, 137; *Tess of the d'Urbervilles*, 126
heath-dwellers, 117–18, 123, 127, 133–34, 136, 138, 178n18
heathlands, 14, 39, 115–39, 152
Hensley, Nathan K., 100
herbicides, 149, 151–55
Hildebrand, Jayne, 105, 107, 176n50, 176n54
History of the Countryside, The (Rackham), 127
Homo economicus, 67, 161
horses, 27–28, 60, 63, 64, 69, 70, 92, 96, 98, 110–11, 152. *See also* cattle; livestock
Hovanec, Caroline, 155–56
human-animal hybrids, 64–65, 74, 95–100, 108, 172n38. *See also* characterization: of animals; dehumanization; mermaids
Huxley, Julian, 144, 146
hypocognition, 2

Illustrations of Political Economy (Martineau), 19, 31–37, 38, 40–41, 46–47, 52, 76, 83, 167n34, 169n65
imperialism, 11, 32–33, 34, 42, 61–63, 66, 69, 146, 168n43, 172n40, 173n58. *See also* inequality
Indigenous knowledges. *See* traditional ecological knowledge
industrial novels, 13, 43–46, 54–56, 59, 64, 69, 73–77, 79, 81, 92, 95, 148, 169n3; reception of, 55, 73–75. *See also under names of specific novels and writers*
inequality: economic, 15–16, 21, 24–28, 32, 35, 42–43, 44–45, 47–48, 50–54, 57–59, 61–62, 64–66, 67–68, 71, 74–75, 91, 113, 166n17, 170n14; racial, 11, 21, 32, 34, 59–63, 66, 71, 75, 96, 168n43, 171n28, 172n37, 173n60
insects, 17, 39, 68–69, 71–72, 91, 96, 128, 146, 147, 151, 152, 154. *See also* bees; worms
interdisciplinarity, 6, 9, 139–40, 142–43, 156, 157, 159–62, 163n1, 183n45. *See also* disciplines, academic
intersectionality, 165n13
Iron Law of Wages, 51–52, 170n14
it-narratives, 151, 182n26

Jackson, Zakiyyah Iman, 171n28
Jakobson, Roman, 120
James, Henry, 109
Jane Eyre (Brontë), 4, 164n8
Jevons, William Stanley, 83–84
Jonsson, Frederik Albritton, 22, 33
Jude the Obscure (Hardy), 126

Kim, Claire Jean, 162
King, Amy, 84
King Lear (Shakespeare), 138
Kingsley, Charles, 43; *Alton Locke*, 54, 58–59, 60–61, 64, 74, 75, 173n60; on Martineau, 46

landscape, 14, 18, 32, 35–37, 39, 100–108, 110, 113, 115–16, 117, 119–24, 126, 129–32, 134, 135, 136–38, 146–47, 150, 151, 157, 168n45. *See also* background, as literary device; description, as literary device; ecological literacy; setting
Last of the Curlews (Bodsworth), 158
Latour, Bruno, 28
Lear, Linda, 181n19
Leopold, Aldo, 3, 29, 127, 147, 149–51, 181n24
Levine, Caroline, 9, 163n3, 169n64
Levine, George, 125, 175n32, 177n4
Lewes, George Henry, 78, 81, 82, 85, 86, 112, 163n2
Life of Jesus, Critically Examined, The (Strauss), 81

literary history: defined, 1; of ecology, overview, 1–16; interdisciplinary benefits of, 2–3, 5, 7, 8
livestock, 23–25, 27, 35–36, 44, 56, 58–60, 63, 104–5, 131, 133, 134, 151. *See also* cattle; horses
Lloyd, Edward, 60
Louis Philippe I (king), 42
Lukács, Georg, 102, 104
Lyell, Charles, 66, 76

Macfarlane, Robert, 37
Malay Archipelago, The (Wallace), 71–72
Malthus, Daniel, 20
Malthus, Thomas Robert, 7, 12, 14, 40, 82–83, 91–92; biography of, 20–21, 25, 82; ecological plotting of, 13, 21–22, 24–29, 32, 33, 34–35, 55, 65, 139, 148, 152; *Essay on the Principle of Population*, 12, 19–29, 32, 33, 62, 82, 166n17, 166n20, 167n26; influence of, 12–13, 19–20, 22, 26, 29–30, 44, 65, 143, 148; population theory of, 12, 21, 24–25, 27–28, 32, 33–34, 38, 51–52, 137, 143; *Principles of Political Economy*, 83; reception of, 7, 12, 13, 19–20, 21–22, 25–26, 28, 30–31, 55, 70
Malthusian, as term, 21
"Manchester Strike, A" (Martineau), 47, 51–52
marine life, 32, 33–34, 62, 64, 65, 78, 82, 85–86, 119, 143, 154–55, 174n22. *See also* fisheries
Martineau, Harriet, 7, 13, 14, 167n34; *Autobiography*, 30–31; biography of, 30–33; "Cinnamon and Pearls," 18, 19, 32–33, 63; Darwin and, 18, 19, 37–38, 44; "Demerara," 63; ecological plotting of, 19, 32–37, 38, 47–48, 62–63, 139, 148; *Illustrations of Political Economy*, 19, 31–37, 38, 40–41, 46–47, 52, 76, 83, 167n34, 169n65; influence of, 13, 19–20, 30, 37–38, 41, 43, 45, 46–47; Malthus and, 19–20, 30, 31–32, 33–34, 38, 40, 44; "A Manchester Strike," 47, 51–52; reception of, 31, 37–38, 40–41, 46–47; "Sowers Not Reapers," 35–36; "Weal and Woe in Garveloch," 33–34, 62
Marx, Karl, 42, 165n11
Marxist theory, 8
Mary Barton (Gaskell), 13, 45, 46–54, 55, 56–58, 64–66, 66–69, 70, 73, 74–75
Mayhew, Robert, 166n20
Mehner, John, 152–53
mermaids, 64–66, 67, 172n38, 172n40. *See also* human-animal hybrids
metaphor, 2, 119, 174n14, 178n16; in abolitionist debates, 59, 61–62, 171n31; in Darwin, 132, 144; in ecology, 143, 144, 153, 159; in economics, 159; in Eliot, 86, 108–9, 144, 158, 176n59; in Gaskell, 68; in Hardy, 118, 120–22, 132; in Malthus, 23–24, 26
metonymy, 119–20, 127, 135
Middlemarch (Eliot), 80, 92, 108–9, 158
Mill, James, 31, 70, 83
Mill, John Stuart, 161
Millgate, Michael, 114
Mill on the Floss, The (Eliot), 14, 79–80, 91–92, 95–108, 110, 112
misanthropy, 123–24, 160
modernism, 10, 155–57, 182n33, 182n35
Modern Painters (Ruskin), 105
Morgan, Benjamin, 135
Mori, Masahiro, 125, 178n19
Morton, Timothy, 26, 36–37, 104, 164n4, 166n17, 179n25
Muir, John, 146

Narrative of the Life of Frederick Douglass (Douglass), 59–60
natural history, 1, 9, 40–41, 66–67, 78, 112, 114, 142–43, 172n41; political economy and, 7, 13–15, 18–20, 29, 37–40, 44–45, 66–73, 78–79, 82, 85–86, 139–40, 143, 165n11; religion and, 15, 76; social novel and, 7, 13–15, 18–20, 44–46, 66–70, 78–79, 81–82, 84–86, 87, 114–16, 139–40, 172n43, 173n1, 174n22, 175n32

"Natural History of German Life, The" (Eliot), 13–14, 81–82, 84–85, 94
Natural History of Selborne (White), 84
natural selection, theory of, 20, 38, 70–71, 132
Ngai, Sianne, 125
niche, ecological, 144
Nightwood (Barnes), 157
novel, as genre. *See* bildungsroman; industrial novels; realism; social novel

objectification. *See* commodification; dehumanization
objectivity, 1, 15, 40, 71, 130, 139, 140, 145, 147, 163n1
Oekologie, as term, 39, 112. *See also* ecology
On the Origin of Species (Darwin), 14, 20, 29, 38–39, 40, 44, 112, 115, 128, 133
orangutans, 64, 72, 86

Paradis, James, 130
parataxis, 103, 104
Pavlov, Ivan, 2
personification, 74, 118, 122–26, 129–30, 132, 151, 178n19. *See also* anthropomorphism; characterization
pesticides, 149, 151–55. *See also* DDT poisoning
pest management industry, 147–48, 151–55, 181n19
Physiocrats, 22
Picoult, Jodi, 158
plants. *See* crops; forestry; heathlands; landscape; Scotch fir (Scots pine); trees
plot, as literary device, 3, 8–9, 34, 48, 54–55, 75, 88, 91, 97–98, 101–3, 106, 151–53, 159–62, 174n14. *See also* bildungsroman; ecological plot
political ecology, 28. *See also* ecology
political economy, 5, 9, 12–13, 21, 25, 28, 29, 31, 40–41, 43, 51–52, 58, 72–73, 82–84, 138; natural history and, 7, 13–15, 18–20, 29, 37–40, 44–45, 66–73, 78–79, 82, 85–86, 139–40, 143, 165n11;

social novel and, 7–8, 13–15, 18–20, 29, 30–31, 40–41, 43–46, 52–59, 69–70, 73–74, 78–79, 81–82, 84–85, 138–40, 165n11. *See also* economics; *Illustrations of Political Economy* (Martineau)
Population Problem, The (Carr-Saunders), 143
population theory (Malthusian), 12, 21, 24–25, 27–28, 32, 33–34, 38, 51–52, 137, 143
Principles of Political Economy (Malthus), 83

realism: literary, 9, 13–14, 27, 43–45, 47, 54, 64, 78–80, 81–82, 86, 87, 110, 114–17, 119, 120, 128, 140, 144, 150, 155, 157–59, 166n19, 172n43, 174n22, 175n32, 177n65, 182n33; political, 149–50. *See also* social novel
religion: conservation and, 146; Eliot and, 81, 86, 88, 89–90, 95, 100, 110–11; industrial novels and, 46, 74–76, 79–80, 95, 110; Malthus and, 20–21, 25, 29; natural history and, 15, 76, 84, 142; slavery and, 59; Victorians and, 76–77, 81, 105
Return of the Native, The (Hardy), 14, 115, 116–27, 133, 137
Ricardo, David, 31, 40, 51, 83, 170n14
Richardson, Rebecca, 167n34
Riehl, Wilhelm Heinrich, 81–82, 84
rights: animal, 15, 60, 92–93, 111, 156, 171n28, 172n37; human, 59–61, 63–64, 66, 75–76, 171n28, 172n37, 173n60. *See also* values
Robbins, Lionel, 82
Romanticism, 6, 8, 12, 21, 46, 130, 146
Rousseau, Jean-Jacques, 20
Ruskin, John, 105–6, 108

Sand County Almanac, A (Leopold), 3, 147, 150–51, 181n24
scale, 4, 38, 120, 126, 128–33, 135, 160
scenery. *See* background, as literary device; description, as literary device; landscape; setting

Schleiermacher, Friedrich, 76
Scotch fir (Scots pine), 39, 106, 107, 128–31, 133, 134, 136. See also trees
Scott, Sir Walter, 46
Scovell, Edith Joy, 141
Seaside Studies (Lewes), 78
setting, 14, 33, 35–37, 47–49, 56, 63, 79–80, 100–108, 110, 113, 115–16, 117–18, 120–21, 128, 136, 138, 159–60, 177n65. See also background, as literary device; description, as literary device; ecological literacy; landscape
Shakespeare, William, 138, 142
Shelley, Percy, 12
Silent Spring (Carson), 149, 151–55
"Silver Blaze" (Doyle), 157
slavery, 11, 34, 59–63, 66, 172n40
Smith, Adam, 22, 31, 40, 51, 70
social Darwinism, 8
social novel, 8–9, 12, 13, 46–47, 78–80, 81–82, 84–86, 94, 105, 108–12, 114–16, 152, 155, 158, 159, 177n65, 182n33, 183n45; ecology and, 5, 7, 114–16, 126–30, 133–36, 138–40, 152–53, 159; economics and, 5, 7–8, 18–20, 116, 138–40, 159; natural history and, 7, 13–15, 18–20, 44–46, 66–70, 78–79, 81–82, 84–86, 87, 114–16, 138–39, 172n43, 173n1, 174n22, 175n32; political economy and, 7–8, 13–15, 18–20, 29, 30–31, 40–41, 43–46, 52–59, 69–70, 73–74, 78–79, 81–82, 84–85, 138–40, 165n11; religion and, 74–76. See also realism: literary; *and under names of specific novels and writers*
social-problem novels. See industrial novels
"Sowers Not Reapers" (Martineau), 35–36
Spencer, Herbert, 84
Story of Iceland, The (Elton), 141
Story of Sir Francis Drake, The (Elton), 141
Stowe, Harriet Beecher, 46
Strauss, David Friedrich, 76, 81
Supper at Emmaus (Veronese), 105
Survey of English Literature (Elton), 141
Sybil (Disraeli), 54, 58, 64, 74
sympathy, 14, 52–55, 56, 58, 63–65, 73, 74, 82, 84–86, 87–95, 101–2, 105, 108–10, 123–25, 130, 134, 150, 175n32, 176n59. See also values: social novel and

Tarka the Otter (Williamson), 158
Tess of the d'Urbervilles (Hardy), 126
Thackeray, William, 8
Theory of Political Economy (Jevons), 83–84
thermodynamics, 181n6
Thoreau, Henry David, 6
Through the Looking-Glass (Carroll), 119
Tilley, Christopher, 127
traditional ecological knowledge (TEK), 18, 166n4
trees, 23, 32, 35–36, 39, 71, 103, 107, 118, 119, 129–31, 133, 134–35, 136, 143, 151, 152–54. See also Scotch fir (Scots pine)
Trollope, Anthony, 109–10
Tsing, Anna, 138

Ulysses (Joyce), 157
uncanny valley, 125, 178n19, 179n25. See also characterization: of landscapes; personification

values: bildungsroman and, 97–100; crisis of, 44–45, 56, 58, 66–67, 73–76, 148; ecology and, 29, 63–64, 76, 117, 139–40, 144–45, 146–50, 164n5; economics and, 5; human rights and, 59–61, 63–64; pricing and, 15, 22, 24, 45, 56–58, 65, 69, 72–73, 138, 139, 147–48, 160; social novel and, 52–56, 73–76, 79–80, 84–86, 87–95, 98, 101–2, 105–6, 108–12, 139–40, 150, 155, 157–58, 162. See also rights
Veronese, Paolo, 105

Wallace, Alfred Russel, 70–72, 86, 173n51, 173n53
Wallace, George, 152
Warming, Eugenius, 141
water, as resource, 24, 35–36, 49–50, 92, 101–2, 146–47

Watts, May Theilgaard, 37
"Weal and Woe in Garveloch" (Martineau), 33–34, 62
weather, as plot device, 34, 35–36, 48–51, 62, 71, 152
Wessels, Tom, 37
Westminster Review, 78, 80, 81
Whewell, William, 84, 163n2
White, Gilbert, 84

Wild Animals I Have Known (Seton), 158
Williams, Raymond, 8
Woloch, Alex, 89, 175n29
Wordsworth, William, 146
worms, 132–33, 153–54, 179n37
Worster, Donald, 145
Worthington, John, 30
Wuthering Heights (Brontë), 138
Wynter, Sylvia, 161, 172n37

Recent books in the series
UNDER THE SIGN OF NATURE: EXPLORATIONS IN
ENVIRONMENTAL HUMANITIES

Debra J. Rosenthal and Jason de Lara Molesky, editors • *Cli-Fi and Class: Socioeconomic Justice in Contemporary American Climate Fiction*

James Perrin Warren • *Thoreau's Botany: Thinking and Writing with Plants*

Katherine Cox • *Climate Change and Original Sin: The Moral Ecology of John Milton's Poetry*

Jeremy Chow • *The Queerness of Water: Troubled Ecologies in the Eighteenth Century*

Monica Seger • *Toxic Matters: Narrating Italy's Dioxin*

Taylor A. Eggan • *Unsettling Nature: Ecology, Phenomenology, and the Settler Colonial Imagination*

Samuel Amago • *Basura: Cultures of Waste in Contemporary Spain*

Marco Caracciolo • *Narrating the Mesh: Form and Story in the Anthropocene*

Tom Nurmi • *Magnificent Decay: Melville and Ecology*

Elizabeth Callaway • *Eden's Endemics: Narratives of Biodiversity on Earth and Beyond*

Alicia Carroll • *New Woman Ecologies: From Arts and Crafts to the Great War and Beyond*

Emily McGiffin • *Of Land, Bones, and Money: Toward a South African Ecopoetics*

Elizabeth Hope Chang • *Novel Cultivations: Plants in British Literature of the Global Nineteenth Century*

Christopher Abram • *Evergreen Ash: Ecology and Catastrophe in Old Norse Myth and Literature*

Serenella Iovino, Enrico Cesaretti, and Elena Past, editors • *Italy and the Environmental Humanities: Landscapes, Natures, Ecologies*

Julia E. Daniel • *Building Natures: Modern American Poetry, Landscape Architecture, and City Planning*

Lynn Keller • *Recomposing Ecopoetics: North American Poetry of the Self-Conscious Anthropocene*

Michael P. Branch and Clinton Mohs, editors • *"The Best Read Naturalist": Nature Writings of Ralph Waldo Emerson*

Jesse Oak Taylor • *The Sky of Our Manufacture: The London Fog in British Fiction from Dickens to Woolf*

Eric Gidal • *Ossianic Unconformities: Bardic Poetry in the Industrial Age*

Adam Trexler • *Anthropocene Fictions: The Novel in a Time of Climate Change*

Kate Rigby • *Dancing with Disaster: Environmental Histories, Narratives, and Ethics for Perilous Times*

Byron Caminero-Santangelo • *Different Shades of Green: African Literature, Environmental Justice, and Political Ecology*

Jennifer K. Ladino • *Reclaiming Nostalgia: Longing for Nature in American Literature*

Dan Brayton • *Shakespeare's Ocean: An Ecocritical Exploration*

Scott Hess • *William Wordsworth and the Ecology of Authorship: The Roots of Environmentalism in Nineteenth-Century Culture*

Axel Goodbody and Kate Rigby, editors • *Ecocritical Theory: New European Approaches*

Deborah Bird Rose • *Wild Dog Dreaming: Love and Extinction*

Paula Willoquet-Maricondi, editor • *Framing the World: Explorations in Ecocriticism and Film*

Bonnie Roos and Alex Hunt, editors • *Postcolonial Green: Environmental Politics and World Narratives*

Rinda West • *Out of the Shadow: Ecopsychology, Story, and Encounters with the Land*

Mary Ellen Bellanca • *Daybooks of Discovery: Nature Diaries in Britain, 1770–1870*

John Elder • *Pilgrimage to Vallombrosa: From Vermont to Italy in the Footsteps of George Perkins Marsh*

Alan Williamson • *Westernness: A Meditation*

Kate Rigby • *Topographies of the Sacred: The Poetics of Place in European Romanticism*

Mark Allister, editor • *Eco-Man: New Perspectives on Masculinity and Nature*

Heike Schaefer • *Mary Austin's Regionalism: Reflections on Gender, Genre, and Geography*

Printed in the USA
CPSIA information can be obtained
at www.ICGtesting.com
CBHW021716311024
16728CB00002B/77